GW01564021

ISP
INTERACTIVE
STATISTICAL
PROGRAMS

INSEAD
1/8/87

ISP

INTERACTIVE STATISTICAL PROGRAMS

SPYROS MAKRIDAKIS
INSEAD

ROBERT L. WINKLER
Duke University

WEST PUBLISHING COMPANY

St. Paul   New York   Los Angeles   San Francisco

The Interactive Statistical Programs are COPYRIGHT © 1983
by Spyros Makridakis and Robert L. Winkler

COPYRIGHT © 1986 By WEST PUBLISHING COMPANY
50 West Kellogg Boulevard
P.O. Box 64526
St. Paul, MN 55164-1003

Library of Congress Cataloging in Publication Data

Makridakis, Spyros
  interactive statistical programs.

  Includes index.
    1. Interactive Statistical Programs (Computer programs)
2. Statistics—Computer programs. 3. Mathematical
statistics—Computer programs. I. Winkler, Robert L.
II. Title.
QA276.4M26 1986   519.5′028′5425   84-22065
ISBN 0-314-77919-1

1st Reprint—1986

Contents

Preface

This book is designed to be used in conjunction with a set of interactive statistical programs, called ISP, which has been developed to improve and facilitate the teaching and learning of basic statistics. The ISP software and this book are intended for use in an introductory statistics course. Alternatively, they can be used for self-paced individual learning of statistics, as a ''refresher course'' for those who studied statistics some time ago, and as a convenient computational tool and accompanying reference manual for users of statistics. Introductory statistics courses are seldom among the more popular courses in the curriculum, and we strongly believe that students of statistics can profit from using this book and the ISP software as an integral part of an introductory statistics course.

A very important aspect of ISP and this book is that *no* previous background is needed in either statistics or computers, and the mathematics needed is also minimal (say, the equivalent of the prerequisite for a very low-level basic statistics course). The software is extremely easy to use, is completely interactive and user-friendly, and requires absolutely no programming. ISP facilitates the learning of statistics by removing the computational burden of statistics and enabling students to concentrate on statistical concepts. Moreover, provisions for working with real data (including data chosen by the instructor or student), sampling from data sets, conducting simulations, and experimenting with games of chance help to make statistics more interesting and to give students a better understanding of the nature of uncertainty, sampling variability, and statistical modeling.

The book begins with an introduction to ISP which describes various aspects of the programs and should enable a user to get started with ISP. The rest of the book is organized by statistical concepts. First, it is divided into seven parts, covering descriptive statistics, probability and probability distributions, sampling and sampling distributions, estimation and hypothesis testing, comparing two or more populations, regression and correlation, and forecasting. Each part is then divided into chapters. A chapter begins with a brief introduction of the concepts of interest, gives a description of the ISP commands that are useful for these concepts, provides a solved example to illustrate both the statistical concepts and the ISP commands, and includes a few exercises to help you use ISP and learn the concepts.

For information about purchasing the ISP programs, write to

Lincoln Systems Corporation
P.O. Box 391
Westford MA 01886

or call   (617) 692-3910.

The programs are written in FORTRAN and are currently available for VAX computers and IBM-compatible personal computers. The micro version of ISP consists of three modules which are available separately or as a package:

Main Module—Data handling, Statistical spreadsheet, Transformations, Distributions, Basic statistical functions, Games

Training Module—Descriptions of statistical concepts, Exercises

Correlation/Regression Module—Correlation, Regression, Forecasting

The micro version will run on any IBM-compatible personal computer with 256kb RAM, two 320kb floppy disk drives or one floppy and one hard disk, PC or MS DOS (version 2.0 or greater), and a monochrome or color monitor. The VAX version of ISP covers all of the material in the three micro modules and runs on any Digital VAX with tape drive. For information on possible adaptations for other computers, write or call Lincoln Systems Corporation.

We are grateful to all of the people who provided help and encouragement in bringing this work to completion. Paule Villain did much of the original programming for ISP, and Tim Davidson and Mike Harde were involved in adaptations. Students at INSEAD used ISP in various stages of development and provided valuable feedback, as did some colleagues who taught ISP. INSEAD and Indiana University also provided important support to facilitate the development of ISP and the writing of this book. Finally, Dick Fenton gave helpful advice during the development stage and David Farr did an excellent job of shepherding the manuscript through the production process.

We hope that this book and the ISP software will make the learning of statistics more interesting and more enjoyable and that it will help provide students with a better understanding and appreciation of statistics.

SPYROS MAKRIDAKIS
Fontainebleau, France

ROBERT L. WINKLER
Durham, North Carolina

Introduction to ISP: Interactive Statistical Programs

ISP is a set of Interactive Statistical Programs to help you learn statistics more effectively and with less effort by using a computer terminal or a microcomputer. ISP allows you to start applying from the very beginning the various statistical methods and concepts you are learning. It takes full advantage of today's computer capabilities to facilitate learning by simplifying the data handling and computational chores. The ISP training modules have been developed for people having little or no mathematical background. As you gain experience, you will soon find that ISP's extensive collection of statistical functions is not only valuable pedagogically but extremely useful in a wide variety of practical applications of statistics.

The purpose of this introductory chapter is to give you an idea of the nature and scope of ISP. It should also provide enough information about some basic commands for entering and manipulating data to enable you to start using ISP. Occasionally, a technical term such as "mean" or "standard deviation" is used in this chapter. If you do not understand these terms, don't worry. They will be explained and illustrated later in this book.

THE ISP COMMAND LANGUAGE AND THE MENU-DRIVEN APPROACH

ISP responds to various commands, which are easy to learn and which instruct the computer to do a variety of tasks. A list of these commands can be found in the *ISP Quick Reference Guide,* or you can see them on the screen by typing NAMES after ⟨n⟩, the command prompt. The various commands are grouped in categories (teaching MODULES, STATISTICS, GAMES, TRANSFORMATIONS, etc.), which can be listed at the terminal by entering the first one or two letters of each group of commands. For instance, the letter G entered after ⟨n⟩ will provide you with a list of all the games-related commands.

GAMES and SIMULATIONS

GCARD:	Simulates a deck of playing cards
GCOIN:	Simulates the tossing of coins
GCOMB:	Combinations of M elements taken K at a time
GDIE :	Simulates the throwing of dice
GLOTO:	Simulates the game of lotto
GMARB:	Simulates sampling marbles from an urn
GPERM:	Permutations of M elements taken K at a time
GROUL:	Simulates the game of roulette
GSAMP:	Selects a sample from a population of your choice
GTACK:	Simulates the tossing of a thumbtack

In the menu-driven version of ISP, the programs are even easier to use because you are prompted by a menu, which is a list of alternatives. By entering a letter or moving a cursor on the screen, you choose a command or group of commands. For instance, when you first enter ISP, the main menu, listing the major groups of commands, is shown on the screen.

Interactive Statistical Programs

ENTER THE LETTER OF YOUR CHOICE OR MOVE CURSOR TO YOUR CHOICE

A Introduction to statistics and ISP
B Games and simulations
C Transformations
D Basic statistics
E Calculations
F Distributions
G Plotting and printing
H Regression and analysis of variance
I Forecasting
J Enter, edit, store, and list data/files; modify level/language
X Exit

If you choose the group of commands called BASIC STATISTICS, a menu showing all of the commands in this group will be displayed.

Basic Statistics

ENTER THE LETTER OF YOUR CHOICE OR MOVE CURSOR TO YOUR CHOICE

A Compute the covariance and correlation of two variables
B Compute point and interval estimates

C Compute expected values
D Test hypotheses
E Compute the skewness and kurtosis
F Compute the mean
G Compute the median
H Compute the mode
I Compute the range
J Compute the standard error of the mean
K Compute the variance and standard deviation
L Bayesian statistics
M Summary statistics
R Return to main menu
X Exit

Note that the options include returning to the main menu or exiting from the program.

Occasionally you may find that you have inadvertently chosen the wrong command or answered some question inappropriately while using the command. Many of the commands ask you which column of your data matrix you would like to work on. If you respond with a "0" (zero), the program will exit from the command and let you choose a new command or repeat the old command. As noted above, options such as returning to the main menu or exiting from the programs are available in the menus. To end your session with ISP, you can choose the Exit option or you can type BYE, EXIT, STOP, END, or QUIT.

SELF TEACHING MODULES

There are more than 40 teaching modules available in ISP, including descriptive statistics, probability, sampling, estimation, hypothesis testing, comparison of populations, regression and correlation, and forecasting. These modules explain major statistical concepts and tell you how such concepts can be applied or illustrated through the various ISP commands. In order to use a teaching module you simply need to type its name after $\langle n \rangle$. Refer to the *ISP Quick Reference Guide* for a complete list of the teaching modules; alternatively, type M in ISP to get a list of the modules. To use any of the teaching modules, type the command (such as MDA3) after $\langle n \rangle$.

$\langle n \rangle$ MDA3

***** Measures of Location *****

The mean, median, and mode are three commonly used measures of location. They tell us something about the "typical" member of a data set. The mean is found by adding up all of the members of a data set and then dividing this sum by how many numbers there are in the set. This average is called the arithmetic mean and is widely used in statistics. The command SMEAN can be used to compute the mean of a set of numbers. Alternatively, the command PRINT provides a list of the data set with the sum and the mean.

If the values of a data set are ordered from the smallest to the largest, the median is the middle number in this ordered list of numbers. The command SMEDI can be used to find the median of a set of numbers. Alternatively, the command TSORC can be used to order the data from the smallest to the largest. The median is then the middle value.

The mode is the number that appears most often in a set of data. This number can be found by using the command SMODE. Alternately, the command TSORC can be used. Since the numbers are sorted, those that have the same value will be next to each other and can be counted easily to find the value that occurs with the highest frequency.

Here is an example of the use of the commands SMEAN, SMEDI, and SMODE for a set of 230 ages.

⟨n⟩ SMEAN
***** Computes the Mean *****
Which column(s) do you want to work on ? AGES
Mean of your 230 numbers for:
— Column 1 (AGES) :   27.839130

⟨n⟩ SMEDI
***** Computes the Median *****
Which column(s) do you want to work on ? AGES
Median of your 230 numbers for:
—Column 1 (AGES) :   28.000000

⟨n⟩ SMODE
***** Computes the Mode *****
Which column(s) do you want to work on ? AGES
Mode(s) of your 230 numbers for:
—Column 1 (AGES) :   28.000000

To get some practice with these commands, take a file from those that are stored in the computer (type LISTF for a list of the available files), or enter your own data set, and use SMEAN, SMEDI, and SMODE. Also, try to find the mean, median, and mode without using these commands. (The commands PRINT and TSORC will be helpful here.) By using different data sets and looking at their histograms as well as the three measures of location, you can see how the shape of the histogram affects the mean, median, and mode.

HELP STATEMENTS

With some ISP commands, you are asked if you want help in using the command. A yes answer will tell the computer to provide a brief "HELP" statement concerning the command. Furthermore, even when the computer does not ask if you want assistance, in most instances you can get some clarification or a "HELP" statement by pressing the carriage return key instead of answering a question posed by the computer. The "HELP" statements make it easier to use ISP without having to refer to this book repeatedly during a session with ISP. The book does, however, provide explanations of statistical concepts and examples of the use of ISP commands. The ISP programs are very easy to use, and the book is a useful supplement to the programs.

In terms of assistance, ISP can be used at two levels. While prompts such as "Do you need help?" are often welcome to a beginning user of ISP, a more experienced user may find that having to read and respond to such questions is irritating. Thus, you are given a choice of levels: "beginner" or "experienced user." To change from one level to the other, simply use the command CHANG.

CHOICE OF LANGUAGE

ISP is available not just in English, but also in other languages such as French. Some versions of the ISP programs include more than one language. For such versions, you will be asked which language you would like to use when you first use the programs. If you want to change the language at any time, the command CHANG can be used to change language as well as level.

SAVING RESULTS

Once you have finished with your ISP session, type BYE, EXIT, STOP, END, or QUIT. This will end your session with ISP. However, remember that nothing is lost while using ISP (unless the computer breaks down or you break out of ISP). Everything important appearing on the screen is kept in the file named RESULT.DAT, which is automatically stored in the memory of the computer until you use ISP again. Every time ISP is used, the old contents of RESULT.DAT are destroyed and a new version is created. If you do not want this to happen and you want to keep the contents of the RESULT.DAT file, rename it before using ISP again.

EXERCISES TO REINFORCE LEARNING

There are plenty of training exercises available in ISP that allow you to reinforce learning and to get practical experience using the various statistical concepts. You can use the exercises by typing MEXER after ⟨n⟩. An indication of how well you have done on these exercises will be given to you after you finish each session.

GAMES TO ILLUSTRATE PROBABILITY AND UNCERTAINTY

A good deal of statistics involves learning about probabilities and understanding uncertainty. This learning can be facilitated by the various games of chance available in ISP. For a list of the games available, see the *ISP Quick Reference Guide,* or type G after ⟨n⟩. In order to play any of the games, type its name after the ⟨n⟩ prompt.

In order to use the modules, the exercises, or the games, you do not need to enter any of the data on your own. Any required information is supplied in the programs. You only need to answer the various questions asked by the programs. These will be adequate to get results or to achieve some learning objective. However, if you do want to work with your own numbers you will first have to enter your data in the computer. (See the discussion of datafile operations later in this chapter.)

COMPUTING WITH ISP AS A HAND CALCULATOR

In some cases you might wish to do arithmetic operations on data. For instance, you might want to do some calculations such as

$$(35 + 4(29.3)/6)^2 + \sqrt{29 + 126} \quad .$$

You can find the value of this expression by using ISP's CALC command.

⟨n⟩ CALC
           * * * Hand Calculator * * *

Enter the mathematical expression to be computed :
$(35 + 4*29.3/6)^2 + (29 + 126)^{.5}$

The result of your computations is : 2986.33430

Do you have more computations ? N

If you have some simple calculations to do, you can use any of the C (computing) commands, which allow you to use ISP as a powerful hand calculator. For instance, squaring five numbers and adding the results can be done as follows:

⟨n⟩ CSXX

       ***** Sums n Squared Numbers *****

Enter the n numbers :
12.3 25 32.12 67 25.5

The sum of your   5 squared numbers is 6947.2344

The computer does not store these data or remember them for subsequent uses.

ISP AS A STATISTICAL SPREADSHEET

ISP provides the statistics student or practitioner the ability to create a multiple-column worksheet, using typical transformations on a column-by-column basis. The ISP transformations include arithmetic operations, differencing, standardizing, and sorting. Here is an example:

⟨n⟩ TCC

***** Squares all elements of a Column *****

Observ.	Value	(This is now column no: 2)
1	729.00000	
2	625.00000	
3	625.00000	
4	1024.0000	
5	900.00000	
6	841.00000	
.	.	
.	.	
.	.	
225	841.00000	
226	841.00000	
227	625.00000	
228	1369.0000	
229	1156.0000	
230	841.00000	
Sum	180113.00	
n =	230	
Mean	783.10000	

Give a title for variable no. 2: XSQU

⟨n⟩ PRINT

* * * To Print Data * * *

Do you want to print:
1 . The first 8 and last 8 observations
2 . Rows specified by you
3 . Successive rows
4 . All rows
Enter a number from 1 to 4: 1

Observ.		AGES	XSQU
*	1*	27.000	729.000
*	2*	25.000	625.000
*	3*	25.000	625.000

*	4*	32.000	1024.000
*	5*	30.000	900.000
*	6*	29.000	841.000
*	7*	26.000	676.000
*	8*	28.000	784.000
	.	.	.
	.		
	.	.	.
*	223*	34.000	1156.000
*	224*	34.000	1156.000
*	225*	29.000	841.000
*	226*	29.000	841.000
*	227*	25.000	625.000
*	228*	37.000	1369.000
*	229*	34.000	1156.000
*	230*	29.000	841.000
Sum		6403.0000	180113.0000
n =		230	230
Mean		27.8391	783.1000

ISP'S STATISTICAL FUNCTIONS

Many commands are provided to allow you to use all of the concepts learned in the teaching modules on the statistical spreadsheet data. These include correlation, regression, confidence intervals, hypothesis testing, and many descriptive statistics (mean, mode, median, standard deviation, standard error, range, variance, skewness, kurtosis, etc.). In addition, several important probability distributions are provided.

PLOTTING AND PRINTING DATA

Various ISP commands are available to print or plot your data at the terminal. They range from listing data for editing purposes (using PRINT and EDIT) to illustrative graphics (such as histograms, scatter plots, time series plots, and plots of data and theoretical distributions). For instance, to print your data, type PRINT after ⟨n⟩.

⟨n⟩ PRINT

* * * To Print Data * * *

Observ.		X1	X2	Y
*	1*	12.000	30.000	100.500
*	2*	15.000	43.000	124.000
*	3*	14.000	35.000	114.300

*	4*	22.000	63.000	205.000
*	5*	19.500	38.700	127.100
*	6*	33.200	52.000	301.000
*	7*	26.000	75.000	251.000
*	8*	25.000	70.000	209.000
*	9*	44.000	131.000	438.000
*	10*	29.000	73.000	333.700
Sum		239.7000	610.7000	2203.6000
n =		10	10	10
Mean		23.9700	61.0700	220.3600

DATAFILE OPERATIONS

There are many practical applications that require data to be used in several different statistical operations. For instance, you might want to calculate the mean and the variance of the data, and later on you might decide that you want the median and the mode of the same data. It would be impractical and time consuming if you had to input your data anew each time you wanted to compute different statistical measures. To avoid unnecessary effort you can enter your data once and keep them in the memory of the computer until you finish your session. In order to do this you must enter your data by the ISP commands ENTER or RENTE. The command ENTER can be used as follows to create a one-variable datafile:

⟨n⟩ ENTER

     * * * To Enter a Data Set * * *

Need help ? N

Do you want to input your data:
1 . From the terminal
2 . From a file

Enter 1 or 2: 1

Enter n numbers:
123,105.23 88 129 100,120.1234 99.23,120,115 104

Observ.	Value	(This is now column no: 1)
1	123.00000	
2	105.23000	
3	88.000000	
4	129.00000	
5	100.00000	
6	120.12340	
7	99.230000	
8	120.00000	
9	115.00000	
10	104.00000	
Sum	1103.5834	
n =	10	
Mean	110.35834	

Give a title for variable no. 1: X

When the numbers are entered, they must be separated by either a space or a comma so that the computer knows when one number ends and another begins. In the example, spaces are used in some instances and commas in other instances.

The command RENTE can be used as follows to create a datafile containing several variables.

⟨n⟩ RENTE

* * * To Enter a Multivariate Data Set * * *

Need help ? N

Do you want to input your data:
1 . From the terminal
2 . From a file

Enter 1 or 2: 1

Number of observations ? 10

Number of variables ? 3

For observation   1 Enter   3 variables
12 30 100.5

For observation   2 Enter   3 variables
15 43 124

For observation   3 Enter   3 variables
14 35 114.3

For observation   4 Enter   3 variables
22 63 205

For observation   5 Enter   3 variables
19.5 38.7 127.1

For observation   6 Enter   3 variables
33.2 52 301

For observation   7 Enter   3 variables
26 75 251

For observation   8 Enter   3 variables
25 70 209

For observation   9 Enter   3 variables
44 131 438

For observation   10 Enter   3 variables

29 73 333.7

Your new variables have been stored from column 1
to column   3.

Give a title for variable no.   1: X1

Give a title for variable no.   2: X2

Give a title for variable no.   3: Y

Go through command PRINT to see the contents of your data matrix.

Once the data set has been entered through ENTER or RENTE, it is stored in the memory of the computer until you have finished your session with the computer. If you do not want to use these data again, you do not need to do anything more. Often, however, you might want to keep your data for later sessions. Next we discuss how this can be done.

STORING DATA FOR LATER USE

There are times when you will have a lot of data that you want to use on different days, so you should permanently store your data in a computer file. A file is the equivalent of writing the data in a notebook, which you can go back and read as many times as you want. The computer can read files the same way that you can read notebooks, and it treats the data each time as if you had just entered it at the terminal keyboard.

Most computers have file editing programs that allow you to create files. If you know how to use the local editor, you can use the file created by that editor in ISP. If you do not know how to create files and you still want to store your data permanently, enter them in ISP by using the command ENTER or RENTE. Then you can store your data by using the ISP command FSTOR.

⟨n⟩ FSTOR

        ∗ ∗ ∗ To Store Data in a File ∗ ∗ ∗

Need help ? N

File name for storing the data ? TEST

Do you want to store all your data ? Y

⟨n⟩ BYE

You can get a list of the files that are stored by using the command LISTF.

Any time you would like to recall a stored datafile and reuse the data in ISP, you can do so by using ENTER for a data file with a single variable or RENTE for a datafile with two or more variables. Just tell ISP that you want to input data from a file, and then name the file that holds the data. Here is an example involving a file with three variables.

⟨n⟩ RENTE

        ∗ ∗ ∗ To Enter a Multivariate Data Set ∗ ∗ ∗

Need help ? N

Do you want to input your data:
1 . From the terminal
2 . From a file

Enter 1 or 2: 2

Data file name ? TEST

Number of observations ? 10

Number of variables ? 3

Your new variables have been stored from column   1
to column   3.

Go through command PRINT to see the contents of your data matrix.

⟨n⟩ PRINT

        ∗ ∗ ∗ To Print Data ∗ ∗ ∗

Observ.	X1	X2	Y
* 1*	12.000	30.000	100.500
* 2*	15.000	43.000	124.000
* 3*	14.000	35.000	114.300
* 4*	22.000	63.000	205.000
* 5*	19.500	38.700	127.100
* 6*	33.200	52.000	301.000
* 7*	26.000	75.000	251.000
* 8*	25.000	70.000	209.000
* 9*	44.000	131.000	438.000
* 10*	29.000	73.000	333.700
Sum	239.7000	610.7000	2203.6000
n =	10	10	10
Mean	23.9700	61.0700	220.3600

EDITING DATA

Once a data set is entered, it may be necessary to modify it in some way. For example, you may discover an error (say, the number 967 was entered as 167). Or you may obtain more data to add to your data set. If your data matrix is getting very big, you may want to delete variables that you no longer need. To make a simple modification, you do not want to take the time and effort required to re-enter the modified data set. Modifying, or editing, the data set can be handled easily in ISP with the command EDIT. Here is an example of the use of EDIT.

⟨n⟩   PRINT

* * * To Print Data * * *

Observ.	SCORE	
* 1*	113.000	
* 2*	115.000	
* 3*	134.000	
* 4*	107.000	
* 5*	142.000	
* 6*	134.000	A value of 941 has
* 7*	941.000	← been entered erroneously.
* 8*	142.000	It should be 141.
* 9*	124.000	
* 10*	118.000	
* 11*	129.000	
* 12*	136.000	
* 13*	124.000	
Sum	2459.0000	
n =	13	
Mean	189.1538	

⟨n⟩   EDIT

* * * To Edit (Modify,Add,Delete) Data * * *

Which operation do you want to do on your data matrix:
1. Modify existing elements
2. Add observations or variables
3. Delete observations or variables
4. Move one column into another one
Enter 1,2,3, or 4 ? 1
Enter line number of the element to modify ? 7
Enter column number of the element to modify ? 1
Old value : 941.00000   ; New value ? 141

More elements to modify ? N
Use the PRINT command to see your new matrix.

⟨n⟩   PRINT

* * * To Print Data * * *

Observ.	Score
* 1*	113.000
* 2*	115.000
* 3*	134.000
* 4*	107.000
* 5*	142.000
* 6*	134.000
* 7*	141.000
* 8*	142.000
* 9*	124.000
* 10*	118.000
* 11*	129.000
* 12*	136.000
* 13*	124.000
Sum	1659.0000
n =	13
Mean	127.6154

The 941 has been replaced
← by 141, the correct value.

HANDLING MISSING DATA AND OUTLIERS IN ISP

Another interesting aspect of ISP is its TIF? command, which allows you to create index variables, use the equivalent of the conditional IF statement of the computer, and deal with outliers (unusual values).

For instance, if -99.999 is used to denote a missing value, you can use TIF? to substitute the mean for the missing values.

Observ.	SALES	AUTOP	BUILD	
* 1*	280.000	3.910	9.430	
* 2*	281.500	5.120	10.360	
* 3*	337.400	6.650	14.530	
* 4*	403.600	5.020	16.730	
* 5*	402.700	4.620	-99.999	←
* 6*	452.000	6.430	18.520	Missing values have been
* 7*	432.200	5.300	19.630	denoted by -99.999. These
* 8*	582.000	7.850	23.460	missing values can be
* 9*	596.600	5.890	32.710	replaced by the mean or
* 10*	618.800	6.420	-99.999	← any other desired value.
* 11*	513.800	4.460	36.090	
* 12*	607.200	5.670	36.750	
* 13*	629.100	7.250	35.480	
* 14*	602.700	5.540	37.140	
* 15*	656.700	6.930	41.300	
* 16*	775.800	6.740	45.650	
* 17*	877.600	7.650	47.480	
* 18*	893.600	8.000	-99.999	←
* 19*	903.700	6.880	52.170	
* 20*	898.000	6.750	50.060	

⟨n⟩ TIF?

Do you want to:
1. Create a new variable (index)
2. Use certain value(s) of an existing variable to create a new one
3. Replace certain value(s) of a variable by another value
4. Deal with outliers. (answer 1, 2, 3 or 4) ?
  3
Do you want to replace:
1. A single value of the variable
2. A range of values of the variable
3. Everything less than a certain value of the variable
4. Anything larger than a certain value of the variable
Answer 1, 2, 3, or 4:
  1
Enter the single value to be replaced:
-99
Enter the value to replace with:
  31.03

Observ.	SALES	AUTOP	BUILD	
* 1*	280.000	3.910	9.430	
* 2*	281.500	5.120	10.360	
* 3*	337.400	6.650	14.530	
* 4*	403.600	5.020	16.730	
* 5*	402.700	4.620	31.030	← Missing values have been
* 6*	452.000	6.430	18.520	replaced by the mean
* 7*	432.200	5.300	19.630	of the data.
* 8*	582.000	7.850	23.460	
* 9*	596.600	5.890	32.710	
* 10*	618.800	6.420	31.030	←
* 11*	513.800	4.460	36.090	
* 12*	607.200	5.670	36.750	
* 13*	629.100	7.250	35.480	
* 14*	602.700	5.540	37.140	
* 15*	656.700	6.930	41.300	
* 16*	775.800	6.740	45.650	
* 17*	877.600	7.650	47.480	
* 18*	893.600	8.000	31.030	←
* 19*	903.700	6.880	52.170	
* 20*	898.000	6.750	50.060	

If there are outliers ("unusual" values), you could "clip" them (i.e., replace them by the value of the mean plus or minus a certain number of standard deviations). The mean and standard deviation used in this operation are calculated with the outliers omitted.

Observ.	X	
* 1*	3.000	
* 2*	5.000	
* 3*	7.000	
* 4*	4.000	
* 5*	6.000	
* 6*	8.000	
* 7*	3.000	
* 8*	2.000	
* 9*	7.000	
* 10*	2.000	
* 11*	4.000	
* 12*	3.000	
* 13*	7.000	
* 14*	6.000	Outlier. Its value can be
* 15*	33.330	← dealt with using any of the three options given in TIF? for
* 16*	6.000	handling outliers.
* 17*	2.000	
* 18*	7.000	
* 19*	9.000	
* 20*	2.000	
Sum	126.3300	
n =	20	
Mean	6.3165	

⟨n⟩ TIF?

Do you want to:
1. Create a new variable (index).
2. Use certain value(s) of an existing variable to create a new one
3. Replace certain value(s) of a variable by another value
4. Deal with outliers. (answer 1, 2, 3, or 4)?
   4
Do you want to:
1. Eliminate the outliers
2. Replace (reduce/clip) the outliers
3. Replace the outliers by the mean (answer 1, 2, or 3)
   2
How many standard deviations away from the mean do you want to set the outliers.
(I suggest you answer 3)?
   3.000000

1 outliers of variable 1 were reduced to the value of the mean plus/minus
3.0 standard deviations.
The value of the mean was calculated without the outliers.

Observ.	X	
* 1*	3.000	
* 2*	5.000	
* 3*	7.000	
* 4*	4.000	
* 5*	6.000	
* 6*	8.000	
* 7*	3.000	
* 8*	2.000	
* 9*	7.000	
* 10*	2.000	
* 11*	4.000	
* 12*	3.000	Outlier has been changed to
* 13*	7.000	the mean + 3 times the
* 14*	6.000	standard deviation of the
* 15*	11.743	← data. Notice that the outlier
* 16*	6.000	has been excluded while
* 17*	2.000	calculating the mean and
* 18*	7.000	standard deviation.
* 19*	9.000	
* 20*	2.000	
Sum	104.7427	
n =	20	
Mean	5.2371	

Finally, you can create an index variable, such as the year, as can be seen below.

⟨n⟩ TIF?

Do you want to:
1. Create a new variable (index)
2. Use certain value(s) of an existing variable to create a new one
3. Replace certain value(s) of an existing variable by another value
4. Deal with outliers. (answer 1, 2, 3, or 4)?
   1
Give the first value of the variable (index) you want to create?
1963
Enter a desired increment for successive values:
   1

Observ.	Value	
1	1963.0000	(This is now column no: 2)
2	1964.0000	
3	1965.0000	
4	1966.0000	
5	1967.0000	
6	1968.0000	
.	.	
.	.	The index variable YEAR was
.	.	created.
15	1977.0000	
16	1978.0000	
17	1979.0000	
18	1980.0000	
19	1981.0000	
20	1982.0000	
Sum	39450.000	
n =	20	
Mean	1972.5000	

Variable :   2 (YEAR)
Column created was :   2 YEAR

SCIENTIFIC NOTATION

When numbers get quite large or quite small, they may be printed by the computer in scientific notation. In this notation, a number is expressed as a value multiplied by some power of ten. For example,

$$0.64786963 \; E + 09 \text{ is } (0.64786963) \, (10^9) \quad ,$$

$$\text{or } 647,869,630 \quad .$$

The "E + 09" indicates that the decimal point should be moved nine places to the right. A negative exponent tells us to move the decimal place to the left. For instance,

$$0.64786963 \text{ E} - 07 \text{ is } (0.64786963)(10^{-7}) \quad ,$$

$$\text{or } 0.000000064786963.$$

EXAMPLES OF THE USE OF ISP COMMANDS

Several examples involving the various facets of ISP are shown below.

⟨n⟩ ENTER

       * * * To Enter a Data Set * * *

Need help ? N

Do you want to input your data:
1. From the terminal
2. From a file

Enter 1 or 2: 1

Enter n numbers:
12 10.3 21 18 20 17 16 11.2

Observ.	Value	
1	12.000000	(This is now column no: 1)
2	10.300000	
3	21.000000	
4	18.000000	
5	20.000000	
6	17.000000	
7	16.000000	
8	11.200000	
Sum	125.50000	
n =	8	
Mean	15.687500	

Give a title for variable NO. 1: VALUE

⟨n⟩ SVAR

      ***** Computes the Variance and Standard Deviation *****

Here are the results for your   8 numbers:

	Variance	Standard deviation
Column 1 (VALUE) :	16.678393	4.0839188

These are the sample var. and stand. dev. (i.e., the divisor is n-1)

⟨n⟩ ENTER

***** To Enter a Data Set *****

Do you want to erase the old content of your   8 numbers ? N

OK, you will enter the data in column: 2.

Do you want to input your data:
1. From the terminal
2. From a file

Enter 1 or 2: 1

Enter 8 numbers:
120 118.3 110.5 124 95.9 90.4 101.1 98.7

Observ.	Value	(This is now column no: 2)
1	120.00000	
2	118.30000	
3	110.50000	
4	124.00000	
5	95.900000	
6	90.400000	
7	101.10000	
8	98.700000	
Sum	858.90000	
n =	8	
Mean	107.36250	

Give a title for variable NO. 2: QUANT

⟨n⟩ SRANG

***** Computes the range of n numbers *****

Which column(s) do you want to work on ? VALUE QUANT

Here are the results of your 8 numbers : Maximum   Minimum   Range

	Maximum	Minimum	Range
Column 1 (VALUE) :	21.00	10.30	10.70
Column 2 (QUANT) :	124.00	90.40	33.60

⟨n⟩ GCOIN

***** Simulates the Tossing of Coins *****

Need help? Y

I can simulate for you the tossing of coins as many times as you wish. This will allow you to understand how the distribution of heads (or tails) behaves experimentally and to compare that with the theoretical values from the corresponding binomial distribution.

Number of coins (max. 16) : 4

Number of tosses per coin : 20

Toss	Results	Number of Heads
1	T T H T	1
2	T H T T	1
3	H H T H	3
4	T H T T	1
5	T T T T	0
6	T T H T	1
7	H T H T	2
8	H T H T	2
9	H T H T	2
10	T T H H	2
11	H H T T	2
12	T T H H	2
13	H T T T	1
14	H T H H	3
15	T H H T	2
16	H H T T	2
17	H H T H	3
18	H H T H	3
19	H H H T	3
20	H H H T	3

39

Do you want a frequency distribution of the number
of HEADS per toss : Y

Number of Heads	Abso. Freq.	Relative Freq.	Theoretical Freq.
0	1	.050	.063
1	5	.250	.250
2	8	.400	.375
3	6	.300	.250
4	0	.000	.063
	20	1.000	1.000

⟨n⟩ DNORM

***** Normal Distribution *****

Need help ? N

Do you want :
1 . A table of normal probabilities
2 . A single normal probability
3 . The value of X corresponding to some cumulative prob.

Enter 1,2, or 3 ? 1

Areas to the right of z value

z	.00	.01	.02	.03	.04	.05	.06	.07	.08	.09
0.0	.5000	.4960	.4920	.4880	.4840	.4801	.4761	.4721	.4681	.4641
.1	.4602	.4562	.4522	.4483	.4443	.4404	.4364	.4325	.4286	.4247
.2	.4207	.4168	.4129	.4090	.4052	.4013	.3974	.3936	.3897	.3859
.3	.3821	.3783	.3745	.3707	.3669	.3632	.3594	.3557	.3520	.3483
.4	.3446	.3409	.3372	.3336	.3300	.3264	.3228	.3192	.3156	.3121
.5	.3085	.3050	.3015	.2981	.2946	.2912	.2877	.2843	.2810	.2776
.6	.2743	.2709	.2676	.2643	.2611	.2578	.2546	.2514	.2483	.2451
.7	.2420	.2389	.2358	.2327	.2296	.2266	.2236	.2206	.2177	.2148
.8	.2119	.2090	.2061	.2033	.2005	.1977	.1949	.1922	.1894	.1817
.9	.1841	.1814	.1788	.1762	.1736	.1711	.1685	.1660	.1635	.1611
1.0	.1587	.1562	.1539	.1515	.1492	.1469	.1446	.1423	.1401	.1379
1.1	.1357	.1335	.1314	.1292	.1271	.1251	.1230	.1210	.1190	.1170
1.2	.1151	.1131	.1112	.1093	.1075	.1056	.1038	.1020	.1003	.0985
1.3	.0968	.0951	.0934	.0918	.0901	.0885	.0869	.0853	.0838	.0823
1.4	.0808	.0793	.0778	.0764	.0749	.0735	.0721	.0708	.0694	.0681
1.5	.0668	.0655	.0643	.0630	.0618	.0606	.0594	.0582	.0571	.0559
1.6	.0548	.0537	.0526	.0516	.0505	.0495	.0485	.0475	.0465	.0455
1.7	.0446	.0436	.0427	.0418	.0409	.0401	.0392	.0384	.0375	.0367
1.8	.0359	.0351	.0344	.0336	.0329	.0322	.0314	.0307	.0301	.0294
1.9	.0287	.0281	.0274	.0268	.0262	.0256	.0250	.0244	.0239	.0233

2.0	.0228	.0222	.0217	.0212	.0207	.0202	.0197	.0192	.0188	.0183
2.1	.0179	.0174	.0170	.0166	.0162	.0158	.0154	.0150	.0146	.0143
2.2	.0139	.0136	.0132	.0129	.0125	.0122	.0119	.0116	.0113	.0110
2.3	.0107	.0104	.0102	.0099	.0096	.0094	.0091	.0089	.0087	.0084
2.4	.0082	.0080	.0078	.0075	.0073	.0071	.0069	.0068	.0066	.0064
2.5	.0062	.0060	.0059	.0057	.0055	.0054	.0052	.0051	.0049	.0049
2.6	.0047	.0045	.0044	.0043	.0041	.0040	.0039	.0038	.0037	.0036
2.7	.0035	.0034	.0033	.0032	.0031	.0030	.0029	.0028	.0027	.0026
2.8	.0026	.0025	.0024	.0023	.0023	.0022	.0021	.0021	.0020	.0019
2.9	.0019	.0018	.0018	.0017	.0016	.0016	.0015	.0015	.0014	.0014
3.0	.0013	.0013	.0013	.0012	.0012	.0011	.0011	.0011	.0010	.0010

⟨n⟩ CDIV

***** Divides the First Number by the Second *****

Enter your 2 numbers: 180113 230

180113.00 / 230.00000 = 783.10000

⟨n⟩ GLOTO

***** Simulates the Game of LOTTO *****

Need help? Y

Lotto is one of the most popular lotteries in France. About 140 million francs are bet every week. I am simulating lotto in every detail. You can play it in any of its variations by choosing any of the options that exist. You can play between 2 French Francs and 210 French Francs and can choose any of the options of grids and numbers. Enjoy yourself and try to figure out your chances of winning. Do not be upset if you do not win too often. I am simply simulating the reality of lotto.

Do you want to play:
      6 numbers for 2, 4, 6, or 8 Francs
      7 numbers for 7 Francs
      8 numbers for 28 Francs
      9 numbers for 84 Francs
      10 numbers for 210 Francs
      Enter 6, 7, 8, 9, or 10: 8

Enter your 8 numbers: 5 23 36 38 39 44 45 49

Your 8 LOTO numbers are: 5 23 36 38 39 44 45 49

Please don't go away. You forgot to pay me! You owe me 28 Francs

The BOWL is spinning! the 6 winning numbers are:

15 23 19 47 6 22 The complementary number is 14

I am sorry you did not win anything. I hope you will be luckier next time.

Do you want some statistics about the game? Y

This week, 134914780 Francs were played.

11049520 FF were distributed at rank no 1 to	8 winners.
5524760 FF were distributed at rank no 2 to	49 winners.
16574280 FF were distributed at rank no 3 to	2478 winners.
16574280 FF were distributed at rank no 4 to	130697 winners.
23940627 FF were distributed at rank no 5 to	2386023 winners.

SUMMARY

As you can see by this introduction to the ISP programs, serious topics of statistics can be introduced in a friendly and entertaining manner. ISP is an excellent device for a refresher course as well as for the first-time statistics student. ISP can be used as a workshop tool to complement class work. It can also be used in a self-study mode. Finally, ISP is an efficient and easy-to-use set of programs to do a wide range of actual statistical work.

A few basic commands for handling data should enable you to start using ISP. You can enter data at the terminal or recall data from a file via the commands ENTER (for a single variable) or RENTE (for several variables). The command PRINT will display your data on the screen, EDIT enables you to edit the data, and TIF? provides options for handling missing data and outliers as well as for creating index variables. Data can be stored in a file by using FSTOR, and the command LISTF will give you a list of the stored files. To change the level ("beginner" or "experienced user") or the language in multilevel, multilanguage versions of ISP, use CHANG. Finally, to end your session with ISP, type BYE, EXIT, STOP, END, or QUIT. The wide variety of other ISP commands will be discussed in the remainder of this book, and you can consult the *ISP Quick Reference Guide* for a list of the ISP commands. For additional introductory material on the ISP programs and on statistics, use the commands MISP and MINTR to see the introductory teaching modules.

The remainder of this book is organized in the same fashion as the teaching modules of ISP. Each chapter corresponds to a module and describes the statistical concepts from that module in a little more detail. Computational aspects are discussed briefly, and we indicate which ISP commands are relevant for the concepts in each chapter. A "solved example" then illustrates the concepts and their application, generally with some ISP output to demonstrate the use of the ISP commands. Exercises are included to provide you with some practice in applying the statistical concepts and in using the ISP commands yourself.

The chapters in this book are arranged into seven groups, or parts of the book:

Part A — Descriptive statistics
Part B — Probability and probability distributions
Part C — Sampling and sampling distributions
Part D — Estimation and hypothesis testing
Part E — Comparing two populations
Part F — Regression and correlation
Part G — Forecasting

At the end of each part a brief summary is presented, along with a vocabulary list, a list of symbols, a list of formulas, and a review of relevant ISP commands.

Exercises

INT.1.

In order to get acquainted with ISP, look over the *ISP Quick Reference Guide* and use the command MISP for a module that provides an introductory session to ISP.

INT.2.

Try some of the computing commands in ISP. For example, use CSQRT to find the square root of 61, use CPOW to raise 28 to the fourth power, and use CALC to find $37(16)(321 - 153)/13$.

INT.3.

Use the command ENTER to enter the 6 numbers 82, 47, 61, 58, 89, and 53. Use the title SCORE for this data set. Then use PRINT to print the data set.

INT.4.

Enter the data from Exercise INT.3. After the six numbers have been entered, use EDIT to change the third number from 61 to 68.

INT.5.

Use ENTER to enter the four numbers 3,6,2, and 5. Then use TCC to square all of these numbers and use PRINT to display the original numbers and their squares.

INT.6.

Try the commands GLOTO, GROUL, or GCARD to play some of the games available in ISP.

PART A
Descriptive Statistics

Large volumes of data are available today. This is partly due to the computer, which has made the storage and manipulation of large quantities of data economical, and partly because of various specialized agencies that have been gathering data over long periods of time. Having the data is not enough, however; something must be done so that these data can be used to help us make better decisions.

Statistics is, indeed, concerned with ways of making sense out of data. Part A of this book will be concerned with ways of describing and summarizing data. The words "describing" and "summarizing" are used interchangeably. The aim of descriptive methods is to help us analyze and understand numerical information (that is, data). This objective becomes very important when large volumes of data are involved, since the information obtained from these data is needed to make better, more intelligent decisions. This is what we mean by making sense out of data: How can we use the information they contain to help us in decision making?

There are various ways of describing or summarizing data. Data can be expressed in terms of frequency distributions, which involve counting how many times certain values or classes of values appear in a data set. The topic of frequency distributions and their graphical presentation in histogram form are dealt with in Chapters A1 and A2. Another type of descriptive measure is a summary measure calculated from the data. Chapters A3 and A4 are concerned with various summary measures such as the mean and standard deviation. Chapter A5 examines how summary measures can be computed, not from raw data but from grouped data or frequency distributions. Then, Chapter A6 discusses how we can attempt to approximate a frequency distribution with just two summary measures, the mean and the standard deviation (which play an extremely useful role in the field of statistics.) Finally, Chapter A7 explains how data can be standardized, a process which is used quite often in statistics.

A1
Frequency Distributions

A frequency distribution is a way of describing or summarizing large volumes of data by grouping them into a limited number of classes or categories. The frequency distribution gives the number of times each class occurs in the data (for example, the number of drivers with 2 accidents in the past year). Alternatives to this absolute frequency distribution are relative frequency distributions (the proportion of drivers with 2 accidents), cumulative absolute frequency distributions (the number of drivers with 2 or fewer accidents), and cumulative relative frequency distributions (the proportion of people with 2 or fewer accidents). These distributions can be found by simply counting the number of times that members of each class occur and making appropriate adjustments: dividing by the total number of data points for relative frequencies, or accumulating for cumulative frequencies.

It is worthwhile to think carefully about how to group the data into various classes. For effective summarization and communication, you should not to have too few or too many classes (some writers suggest aiming for between 5 and 16 classes). The number of classes can be controlled by choosing the class interval width (the difference between the largest and smallest values in a class) and the midpoint (the value midway between these extreme values).

ISP COMMANDS

Once a set of data has been entered into ISP, either from the terminal or from a file, frequency distributions can be generated by using the command PFREQ. PFREQ will provide all four types of frequency distributions:

1. an absolute frequency distribution,
2. a relative frequency distribution,

3. a cumulative absolute frequency distribution, and

4. a cumulative relative frequency distribution.

You can specify a class interval width and midpoint. It might be helpful to use the command TSORC to rank the data to find out where the middle and extreme values are located. Also, as we shall see in later chapters, having the computer calculate some summary statistics may be helpful in choosing a class interval width and a midpoint. Alternatively, the computer can automatically choose a class interval width and a midpoint. For this option, just answer with "−1" when the computer asks you to specify a class interval.

SOLVED EXAMPLE

Suppose we want to construct a frequency distribution of the midterm scores of 30 students taking a statistics exam. The raw data are shown in Table A1.1.

The first step in constructing a frequency distribution is to sort the data. This can be seen in Table A1.2, which shows both the actual midterm scores and, next to these, the same scores ranked from the smallest to the largest by using TSORC. We have given the ranked scores the label SRANK. Once the data have been sorted, or ranked, we can easily find the smallest and largest values. In this particular case, the smallest value is 41, and the largest value is 98. The range is, therefore, 98 − 41, or 57. Next we have to decide on the number of classes we want. Suppose we want about 12 classes. If this is the case, the class interval will be 57 divided by 12, which is 4.75, or about 5.

The next thing we need to specify is some central midpoint so that our frequency distribution will start and finish with round numbers. For instance, 70 seems to be a reasonable midpoint. Table A1.3 presents the frequency distribution of scores generated by using PFREQ with a midpoint of 70 and a class interval width of 5.

In order to construct the frequency distribution in Table A1.3 without using PFREQ, we have to count the number of scores in each class. For example, from Table A1.2, there are 2 scores between 40 and 45 (not including 45, which starts the 45-50 class). This number can be seen in Table A1.3 under the absolute frequency column. Then we

TABLE A1.1   Midterm scores of 30 students

54	74	85
59	71	78
43	93	67
69	82	98
62	84	82
69	41	67
64	74	92
82	70	68
96	59	84
79	64	47

TABLE A1.2 Original and ranked midterm scores of 30 students

Observ.		SCORE	SRANK
*	1*	54.000	41.000
*	2*	59.000	43.000
*	3*	43.000	47.000
*	4*	69.000	54.000
*	5*	62.000	59.000
*	6*	69.000	59.000
*	7*	64.000	62.000
*	8*	82.000	64.000
*	9*	96.000	64.000
*	10*	79.000	67.000
*	11*	74.000	67.000
*	12*	71.000	68.000
*	13*	93.000	69.000
*	14*	82.000	69.000
*	15*	84.000	70.000
*	16*	41.000	71.000
*	17*	74.000	74.000
*	18*	70.000	74.000
*	19*	59.000	78.000
*	20*	64.000	79.000
*	21*	85.000	82.000
*	22*	78.000	82.000
*	23*	67.000	82.000
*	24*	98.000	84.000
*	25*	82.000	84.000
*	26*	67.000	85.000
*	27*	92.000	92.000
*	28*	68.000	93.000
*	29*	84.000	96.000
*	30*	47.000	98.000
Sum		2157.0000	2157.0000
n =		30	30
Mean		71.9000	71.9000

count the number of scores between 45 and 50 (50 not included), the number of scores between 50 and 55, and so on. Since the data have been ranked, this is a rather easy task.

To construct the absolute cumulative frequency distribution, we simply add up, or accumulate, the frequencies. Thus, the cumulative frequency is 2 for the class 40–45, 3 (2+1) for the class 45–50, 4 (3+1) for the class 50–55, 6 (4+2) for the class 55–60, and so on. To construct Column 5 in Table A1.3, each value in Column 3 is divided by 30. This gives the relative frequencies. To construct column 6, we divide each value of Column 4 by 30. This gives us the relative cumulative frequencies.

If you do not want to specify a class interval and a midpoint, you can ask the computer to do so. Table A1.4 shows the frequency distributions generated by the computer by using PFREQ and entering -1 for a class interval (thus instructing the computer to select the class interval and midpoint). The only difference between Tables A1.3 and A1.4 is that the computer does not choose 70 as a round midpoint but rather 72, since 72 is closer to the "middle" of the data as indicated by a summary measure known as the mean. The mean and other summary measures will be discussed in Chapters A3 and A4.

TABLE A1.3 Frequency distribution of midterm scores from PFREQ when a class interval of 5 and a midpoint of 70 are specified

Classes of elements		Absolute freq.	Abs. cumulat. freq.	Relative freq.	Rel. cumulat. freq.
40.000000	44.999999	2	2	0.067	0.067
45.000000	49.999999	1	3	0.033	0.100
50.000000	54.999999	1	4	0.033	0.133
55.000000	59.999999	2	6	0.067	0.200
60.000000	64.999999	3	9	0.100	0.300
65.000000	69.999999	5	14	0.167	0.467
70.000000	74.999999	4	18	0.133	0.600
75.000000	79.999999	2	20	0.067	0.667
80.000000	84.999999	5	25	0.167	0.833
85.000000	89.999999	1	26	0.033	0.867
90.000000	94.999999	2	28	0.067	0.933
95.000000	99.999999	2	30	0.067	1.000
		30		1.000	

TABLE A1.4 Frequency distribution of mid-term scores from PFREQ when the class interval and midpoint are selected by the computer

Classes of elements		Absolute freq.	Abs. cumulat. freq.	Relative freq.	Rel. cumulat. freq.
37.000000	41.999999	1	1	0.033	0.033
42.000000	46.999999	1	2	0.033	0.067
47.000000	51.999999	1	3	0.033	0.100
52.000000	56.999999	1	4	0.033	0.133
57.000000	61.999999	2	6	0.067	0.200
62.000000	66.999999	3	9	0.100	0.300
67.000000	71.999999	7	16	0.233	0.533
72.000000	76.999999	2	18	0.067	0.600
77.000000	81.999999	2	20	0.067	0.667
82.000000	86.999999	6	26	0.200	0.867
87.000000	91.999999	0	26	0.000	0.867
92.000000	96.999999	3	29	0.100	0.967
97.000000	102.000000	1	30	0.033	1.000
		30		1.000	

EXERCISES

A1.1.

The number of accidents reported at a busy intersection during each of the past 20 weeks has been recorded, and the data are given in Table A1.5. Without using the computer, construct a frequency distribution with absolute frequencies for this set of data. Also, construct a relative frequency distribution and cumulative absolute and relative frequency distributions. Then enter the data into ISP and use PFREQ with a class interval width of 1.

TABLE A1.5   Number of accidents reported at a busy intersection for 20 weeks

2	2	3	0	4	3	1	3	5	4
0	1	5	1	2	6	2	1	3	2

A1.2.

Table A1.6 gives the number of children in families living in a small community in France. Use PFREQ to obtain frequency distributions for these data. From the PFREQ output, find the number of families with 3 children, the number of families with 2 or fewer children, and the percentage of families with 4 children.

TABLE A1.6   Number of children in families in small French community

2	3	0	2	8	1	0	4	3	3	1	4	5	4	2	1	3	0	3	2
2	0	3	1	3	2	2	3	1	3	3	3	3	1	2	0	3	2	1	4
5	2	0	1	2	5	4	2	1	1	0	3	5	3	1	2	5	3	2	1
0	0	2	3	1	5	2	9	0	1	4	2	2	4	1	0	2	2	2	1
7	3	0	2	1	2	2	3	2	1	7	3	4	3	1	4	0	2	2	1
7	5	0	0	1	3	2	3	8	1	0	3	0	0	1	3	2	0	2	1
6	3	3	3	1	4	3	2	6	1	2	0	3	6	1	3	4	2	0	1
0	6	2	2	1	3	2	0	4	2	3	2	2	1	0	2	3	2	1	1
6	6	4	2	1	3	2	7	2	1	2	0	4	4	1	2	4	3	0	1
1	4	3	3	0	1	2	2	3	2	0	0	3	3	1	0	2	2	2	1
8	2	0	3	1	2	2	2	4	1	2	2	2	2	1	3	0	2	2	1
7	3	0	0	1	2	2	3	2	1	0	0	3	3	1	2	0	0	2	1
3	4	2	2	1	3	2	2	2	1										

A1.3.

The salaries, in dollars, of 114 managers are given in Table A1.7. Choose a class interval and midpoint and use PFREQ to construct a frequency distribution. Then double the size of the class interval and repeat the process. Finally, use a "−1" to let the computer choose the class interval and midpoint.

TABLE A1.7   Salaries of 114 managers

29400	29400	40565	32890	65000	53656	29827	33616	24360	38976
127500	46375	63540	17100	32656	33406	32886	24472	33129	29232
19828	14280	24360	20726	36540	40194	38976	36540	14040	45099
35017	33748	63540	41412	31668	36540	23862	34104	27022	44064
25578	39536	34104	21924	53000	32155	40040	34591	30000	38976
37758	30000	34104	33409	31668	36520	47502	35808	25000	32886
35322	48008	36712	40000	27500	36540	42951	32886	34975	38976
23862	33955	34104	48720	41412	28200	45100	39372	29232	39884
23385	38976	41168	29232	42360	40595	27720	27500	48720	29475
26827	38001	27000	28512	24360	41479	46020	83300	31500	11388
26796	28560	46020	27370	26796	38976	32123	31104	35700	47600
24353	30450	41412	48720						

A2
Histograms and Graphs

Quite often, there are more effective ways of presenting information than listings of frequency distributions, or collections of numbers in general. Because of the way the human mind works, information can be read and digested more easily if it is in graphical form than if it is a list of numbers. The way we perceive and process pictorial images is much more efficient than the way we interpret collections of numbers.

A histogram is a graphical presentation of a frequency distribution in the form of a "bar chart." It contains exactly the same information as a frequency distribution, except that the frequencies are represented as bars, or rectangles, on a graph. Of course, types of graphs other than histograms are used in statistics (for example, pie charts), but the histogram is the most common of the graphical presentations.

ISP COMMANDS

Once a set of data has been entered into ISP, a histogram of the data can be obtained by using the command PHIST. As in the case of PFREQ, discussed in the preceding chapter, you can specify a class interval width and a midpoint, or you can enter "−1" and let the computer choose these values. The number of classes that can be plotted is restricted to 16, since no more than that number can be fitted into the width of a single page of computer output.

Other useful plot commands are PDIST, which provides a combined plot of data and theoretical distributions, and PROBD, which plots the most common theoretical distributions. These commands will be explained in later chapters when theoretical distributions are discussed.

A computer plotter can be used with PHIST, PDIST, and PROBD to provide graphs that look much nicer than those made by a regular computer printer. A regular computer printer can print characters only in specific spaces (like those of a regular typewriter). In contrast, a graphic terminal or plotter can print lines or figures that look continuous.

SOLVED EXAMPLE

Suppose we want to plot a histogram of the midterm scores given in Table A1.1. Histograms corresponding to Table A1.3 and Table A1.4 are shown in Figures A2.1 and A2.2. For Figure A2.1, PHIST was used with a class interval of 5 and a midpoint of 70. For Figure A2.2, PHIST was used with a response of "−1" for the question about the class interval.

FIGURE A2.1   Histogram of midterm scores from PHIST with class interval of 5 and midpoint of 70

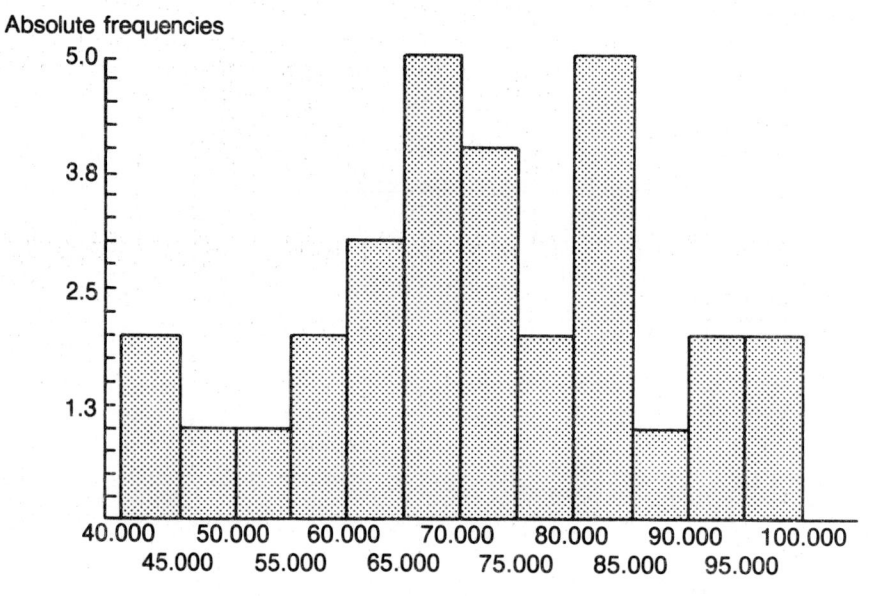

FIGURE A2.2   **Histogram of midterm scores from PHIST with computer choosing the class interval and midpoint**

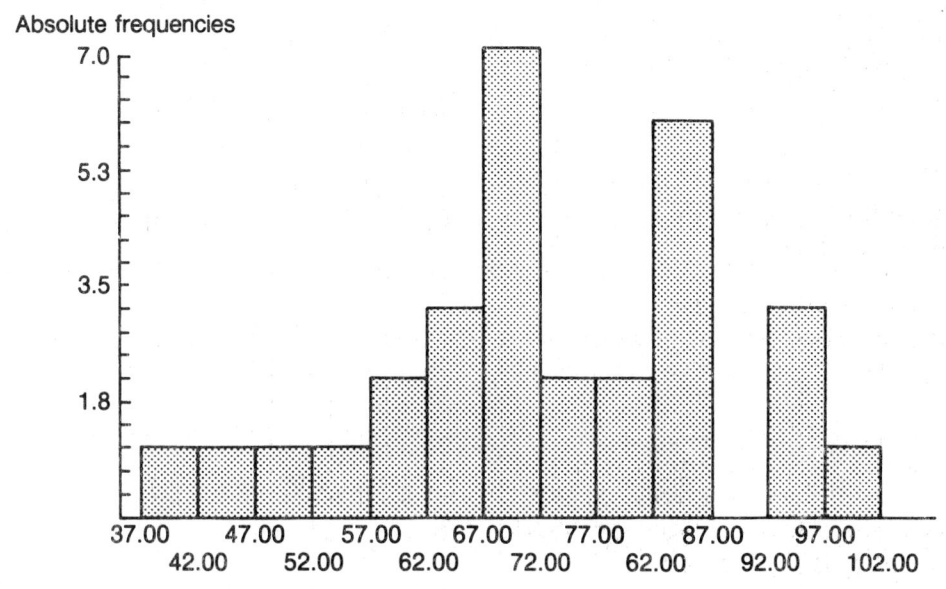

EXERCISES

A2.1.

      Plot a histogram of the data from Exercise A1.1 by hand. Then use PHIST to have ISP generate a histogram for the same set of data.

A2.2.

      Use PHIST to obtain a histogram of the data from Exercise A1.2.

A2.3.

      Choose a class interval and midpoint and use PHIST to find a histogram for the salary data from Exercise A1.3. Then double the size of the class interval and repeat the process. Finally, use a '' − 1'' to let the computer choose the class interval and midpoint.

A3
Measures of Location

In this chapter, we will present three commonly used single summary measures that are referred to as measures of location. A measure of location tells us something about the "typical" member of a data set (for example, a "typical" family income or a "typical" exam grade). In terms of a frequency distribution, a measure of location tells us something about the center of the distribution.

Perhaps the most common measure of location is the mean, which is a simple average found by adding up all of the numbers in a data set and then dividing by how many numbers there are in the set. Another measure of location is the median, which is the middle number in a data set if the data are ordered from the smallest to the largest value. A third measure of location, the mode, is the number that appears most often in a data set.

In statistics, we distinguish between data representing a population (all possible units or observations of interest) and a sample (a subset of the population). Statistical measures such as the mean, median, and mode are called "parameters" when they refer to a population and "statistics" when they refer to a sample. Parameters are often denoted by Greek letters, whereas statistics are usually denoted by Roman letters.For example, the mean of a population of size N with data points $x_1, x_2, \ldots, x_N$ is denoted by μ:

$$\mu = \frac{\sum_{i=1}^{N} x_i}{N} \tag{A3-1}$$

(here Σ denotes summation). The mean of a sample of size n with data points $x_1, x_2, \ldots, x_n$ is denoted by $\bar{x}$ (read x-bar):

$$\bar{x} = \frac{\sum_{i=1}^{n} x_i}{n} \ . \tag{A3-2}$$

ISP COMMANDS

The mean, median, and mode can be computed directly by using the commands SMEAN, SMEDI, and SMODE. However, we advise that these direct commands not be used until you understand how the mean, median, and mode are computed. In general, ISP allows you to find different statistics directly, by specific commands, and to compute them indirectly by some alternative commands.

In the case of the mean, median, and mode, the commands PRINT, TSORC, and PFREQ should help you understand how the computations are done. The command PRINT shows how the mean is computed. The command TSORC sorts and ranks the data from the smallest to the largest. This command can be used to find the median and the mode. The command PFREQ can be used to find the mode; you can also determine the median as the value where 50% of the data are above and 50% are below that value. Thus, in a frequency distribution, the median will be at the point where the cumulative relative frequency changes from less than 0.5 to exactly 0.5 or more than 0.5.

SOLVED EXAMPLE

Summary measures of examination scores are widely used. Suppose we want to compute the mean, median, and mode of the 30 midterm statistics scores shown in Table A1.1.

If we look at Table A1.2, we can see that the mean score is 71.9 (see the last row of the second column). Alternatively, we can use SMEAN, which provides the same number (see Figure A3.1).

FIGURE A3.1   Output from SMEAN, SMEDI, and SMODE for 30 midterm statistics scores

⟨n⟩ SMEAN
***** Computes the Mean *****

Which column(s) do you want to work on ? 1
Mean of your   30 numbers:

— Column 1 (MIDTE) :   71.900000

⟨n⟩ SMEDI
***** Computes the Median *****

Which column(s) do you want to work on ? 1
Median of your   30 numbers:

- Column 1 (MIDTE) :   70.500000

⟨n⟩ SMODE
***** Computes the Mode *****

Which column(s) do you want to work on ? 1
Mode(s) of your   30 numbers:

- Column 1 (MIDTE) :   82.000000

Since there are 30 data points, the median is the value corresponding to observation number

$$\frac{30 + 1}{2} = 15.5$$

when the data points are ordered. But 15.5 is not a whole integer. Thus, to find the median we must find the average of the value corresponding to observation 15 and the value corresponding to observation 16 (see Table A1.2). The median score is therefore

$$\frac{70 + 71}{2} = 70.5 \quad .$$

This median can also be obtained directly by using the command SMEDI, as shown in Figure A3.1.

The mode is 82. The reason for this is that, if we look at the last column of Table A1.2, the value of 82 appears three times and no other value appears more than twice. Thus, the mode of the raw data is 82. Using the command SMODE gives the same result, as the output in Figure A3.1 illustrates.

The histogram of the 30 midterm statistics scores is shown in Figure A3.2. The mean of 71.9 is larger than the median of 70.5 because it is influenced more strongly by

FIGURE A3.2   Histogram of 30 midterm statistics scores

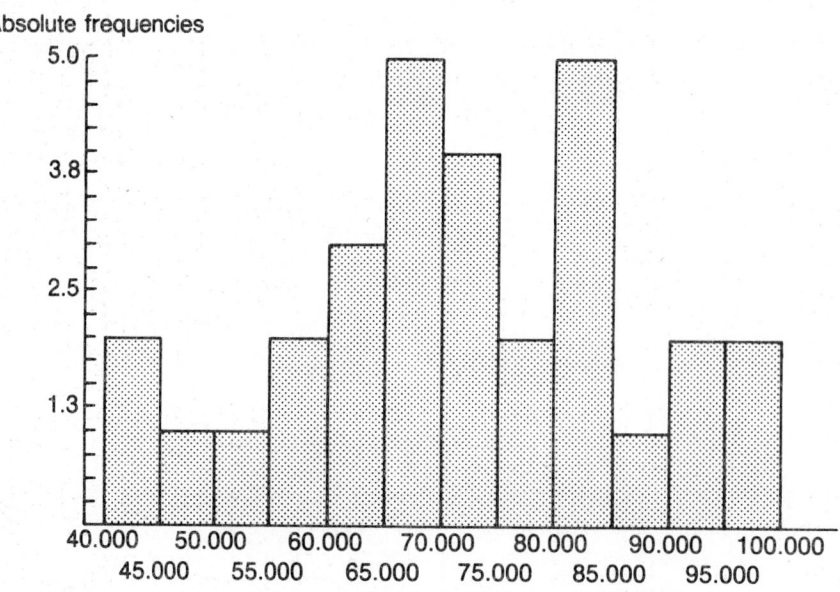

the five scores in the 80–85 range and by the scores in the 90–95 and 95–100 classes than is the median. As for the mode, 82 is the mode of the raw scores, whereas the histogram in Figure A3.2 involves scores grouped into classes of width 5.

EXERCISES

A3.1.

For the data in Table A1.5 concerning the number of accidents at a busy intersection, calculate the mean, median, and mode without using SMEAN, SMEDI, and SMODE. Then use these three commands to let ISP perform the calculations.

A3.2.

Use ISP to find the mean, median, and mode for the data in Table A1.6 regarding the number of children in families in a small French community.

A3.3.

Find the mean, median, and mode of the 114 salaries in Table A1.7, using SMEAN, SMEDI, and SMODE. Compare these three measures of location for this data set and try to explain any differences among the three values.

A4
Measures of Dispersion

Measures of dispersion tell us something about the variation in a set of data. The range, variance, and standard deviation are commonly used measures of dispersion. The range is simply the difference between the largest and the smallest data values. The variance is an average of the squared deviations (differences) between the data values and the mean. The population variance σ^2 is

$$\sigma^2 = \frac{\sum_{i=1}^{N}(x_i - \mu)^2}{N} \ , \tag{A4-1}$$

and the sample variance s^2 is

$$s^2 = \frac{\sum_{i=1}^{n}(x_i - \bar{x})^2}{n-1} \tag{A4-2}$$

(dividing by $n-1$ instead of n is preferable here for reasons related to statistical estimation). For computational purposes, we often use the following alternative formulas for computing variances:

$$\sigma^2 = \frac{N\sum_{i=1}^{N}x_i^2 - \left(\sum_{i=1}^{N}x_i\right)^2}{N^2} \tag{A4-3}$$

and

$$s^2 = \frac{n\sum_{i=1}^{n}x_i^2 - \left(\sum_{i=1}^{n}x_i\right)^2}{n(n-1)} \ . \tag{A4-4}$$

The variance is expressed in squared units. If we take the square root of the variance to avoid squared units, the result is called the standard deviation (σ for the population and s for the sample).

Other types of summary measures are sometimes used. For example, a measure of relative variation is the coefficient of variation, which is simply the standard deviation divided by the mean. A measure of whether the histogram representing the data is symmetric, skewed to the right, or skewed to the left is the coefficient of skewness. The coefficient of kurtosis is a measure of how thin or fat the tails of a distribution are. The most commonly used measures, however, are the measures of location and dispersion presented in this and the previous chapter.

ISP COMMANDS

The range can be computed by ranking the data (command TSORC), printing the ranked data (command PRINT), and subtracting the smallest value from the largest value. Alternatively, the range can be found directly by using the command SRANG.

To compute the variance by formula (A4-1) or (A4-2), first use command TCON to set the second column equal to a constant, the mean; then use TSUB to subtract the second column from the first, putting the result in the third column; and use TCC to square the elements of the third column. The mean of the last column is then the variance if the data represent a population. If the data represent a sample, multiply this result by $n/(n-1)$ to get the sample variance. Another way to find the variance is to use TCC to square the elements of the first column (that is, to square the raw data) and then to use formula (A4–3) or (A4–4). Once the variance is computed, of course, you can find the standard deviation by simply taking the square root of the variance.

The easiest way to find the variance via ISP is to use the command SVAR, which provides both the variance and the standard deviation directly and also gives the coefficient of variation. The divisor used to calculate the variance is the number of observations minus one. If you are dealing with a population, the value that is printed out for the variance should be multiplied by $(N-1)/N$.

The command SUMST provides major summary statistics, including the mean and median as well as the variance, standard deviation, and range. This reduces the need to use multiple commands to find a number of important descriptive measures of location and dispersion.

The coefficients of skewness and kurtosis can be found by using the command SKKU.

SOLVED EXAMPLE

Instructors often provide summary measures of grades or examination scores. Suppose we want to compute the range, variance, and standard deviation of the midterm scores shown in Table A1.1. The range and variance can be computed directly, as can be seen in Figure A4.1, which also shows the mean. Thus, the range of the data is 57, the variance is 218.92, and the standard deviation is 14.796.

FIGURE A4.1 Use of SMEAN, SRANG, and SVAR with statistics scores

$\langle n \rangle$ SMEAN

***** Computes the Mean *****

Mean of your 30 numbers for:
 − Column 1 (X   ): 71.900000

$\langle n \rangle$ SRANG

***** Computes the range of n numbers *****

The range of your 30 numbers is:
Column 1 (X   ): 57.000000   (max value is 98, min value is 41)

$\langle n \rangle$ SVAR

***** Computes the Variance and Standard Deviation *****

These results are the results for your 30 numbers:

Column		Variance	Standard deviation
Column	1 (X):	218.92069	14.795969

To find the range without using SRANG, the largest and smallest values are needed. These can be found by ranking the data with the command TSORC. The ranked data are shown in Table A4.1.

To compute the variance and standard deviation without using SVAR, the commands TSCN (subtracting the mean of 71.9 from the x values) and TCC (squaring the differences) are needed. The results are illustrated in Table A4.2. The sum of the last column is $\sum_{i=1}^{n}(x_i - \bar{x})^2 = 6348.7$. If this is a population, the population variance is 6348.7/30, or 211.62. If it is a sample, 6348.7 must be divided by $n-1$, which is 29, yielding $s^2 = 6348.7/29 = 218.92$.

An alternative way of computing the sample variance is by using formula (A4–4). To do this, we need $\sum_{i=1}^{n} /x_i^2$, which can be found by using TCC to square the column containing the x_i values. The results can be seen in Table A4.3. The sum of column 1 is $\sum_{i=1}^{n} x_i = 2157$, while the sum of column 2 is $\sum_{i=1}^{n} x_i^2 = 161,437$. We can now use formula (A4–4), which yields

$$s^2 = \frac{n\sum_{i=1}^{n} x_i^2 - \left(\sum_{i=1}^{n} x_i\right)^2}{n(n-1)} = \frac{30(161,437) - (2157)^2}{30(29)}$$

$$= \frac{190,461}{870} = 218.92 \quad .$$

TABLE A4.1   Statistics scores ranked from lowest to highest by using TSORC

Observ.		RANKM
*	1*	41.000
*	2*	43.000
*	3*	47.000
*	4*	54.000
*	5*	59.000
*	6*	59.000
*	7*	62.000
*	8*	64.000
*	9*	64.000
*	10*	67.000
*	11*	67.000
*	12*	68.000
*	13*	69.000
*	14*	69.000
*	15*	70.000
*	16*	71.000
*	17*	74.000
*	18*	74.000
*	19*	78.000
*	20*	79.000
*	21*	82.000
*	22*	82.000
*	23*	82.000
*	24*	84.000
*	25*	84.000
*	26*	85.000
*	27*	92.000
*	28*	93.000
*	29*	96.000
*	30*	98.000
Sum		2157.0000
n =		30
Mean		71.9000

Thus, the standard deviation is $\sqrt{218.92}$, which equals 14.796.

EXERCISES

A4.1.

   A sample of five students is taken, and their ages are 23, 21, 26, 25, and 28. Find the range, the sample variance, and the sample standard deviation without using SRANG or SVAR. Then use SRANG and SVAR to find the same summary measures. Finally, use SUMST to obtain summary statistics for this data set.

TABLE A4.2   Computations for finding the variance of 30 statistics scores

⟨n⟩ PRINT

Observ.	X	MEAN	DIFF	SQUAR
* 1*	54.000	71.900	− 17.900	320.410
* 2*	59.000	71.900	− 12.900	166.410
* 3*	43.000	71.900	− 28.900	835.210
* 4*	69.000	71.900	− 2.900	8.410
* 5*	62.000	71.900	− 9.900	98.010
* 6*	69.000	71.900	− 2.900	8.410
* 7*	64.000	71.900	− 7.900	62.410
* 8*	82.000	71.900	10.100	102.010
* 9*	96.000	71.900	24.100	580.810
* 10*	79.000	71.900	7.100	50.410
* 11*	74.000	71.900	2.100	4.410
* 12*	71.000	71.900	− 0.900	0.810
* 13*	93.000	71.900	21.100	445.210
* 14*	82.000	71.900	10.100	102.010
* 15*	84.000	71.900	12.100	146.410
* 16*	41.000	71.900	− 30.900	954.810
* 17*	74.000	71.900	2.100	4.410
* 18*	70.000	71.900	− 1.900	3.610
* 19*	59.000	71.900	− 12.900	166.410
* 20*	64.000	71.900	− 7.900	62.410
* 21*	85.000	71.900	13.100	171.610
* 22*	78.000	71.900	6.100	37.210
* 23*	67.000	71.900	− 4.900	24.010
* 24*	98.000	71.900	26.100	681.210
* 25*	82.000	71.900	10.100	102.010
* 26*	67.000	71.900	− 4.900	24.010
* 27*	92.000	71.900	20.100	404.010
* 28*	68.000	71.900	-3.900	15.210
* 29*	84.000	71.900	12.100	146.410
* 30*	47.000	71.900	− 24.900	620.010
Sum	2157.0000	2156.9999	0.0000	6348.7000
n =	30	30	30	30
Mean	71.9000	71.9000	0.0000	211.6233

A4.2.

Use SRANG and SVAR to compute the range, standard deviation, and variance for the data in Table A1.6 concerning the number of children in families in a small French community.

A4.3.

For the salary data in Table A1.7, find the range, standard deviation, and variance. Also, use SKKU to find the coefficients of skewness and kurtosis.

A4.4.

In Exercise A4.1, suppose the ages were 19, 17, 22, 21, and 24. That is, each age

TABLE A4.3   Some computations for the alternative method of computing the variance of 30 statistics scores

⟨n⟩ PRINT

Observ.		X	XSQU
*	1*	54.000	2916.000
*	2*	59.000	3481.000
*	3*	43.000	1849.000
*	4*	69.000	4761.000
*	5*	62.000	3844.000
*	6*	69.000	4761.000
*	7*	64.000	4096.000
*	8*	82.000	6724.000
*	9*	96.000	9216.000
*	10*	79.000	6241.000
*	11*	74.000	5476.000
*	12*	71.000	5041.000
*	13*	93.000	8649.000
*	14*	82.000	6724.000
*	15*	84.000	7056.000
*	16*	41.000	1681.000
*	17*	74.000	5476.000
*	18*	70.000	4900.000
*	19*	59.000	3481.000
*	20*	64.000	4096.000
*	21*	85.000	7225.000
*	22*	78.000	6084.000
*	23*	67.000	4489.000
*	24*	98.000	9604.000
*	25*	82.000	6724.000
*	26*	67.000	4489.000
*	27*	92.000	8464.000
*	28*	68.000	4624.000
*	29*	84.000	7056.000
*	30*	47.000	2209.000
Sum		2157.0000	161437.0000
n =		30	30
Mean		71.9000	5381.2333

is four years less than in Exercise A4.1. Use ISP to find the range, variance, and standard deviation, and compare these values with the corresponding values found in Exercise A4.1. Repeat the process with the original ages doubled: 46, 42, 52, 50, and 56. Interpret your results.

A5
Grouped Data

Instead of raw data, we sometimes have data that have already been grouped into various classes, with the frequency given for each class. If the *relative* frequency for the class associated with the value x_i is denoted by f_i, then the mean and variance for the grouped data can be found from

$$\mu = \sum_i f_i x_i \qquad \text{(A5–1)}$$

and

$$\sigma^2 = \sum_i f_i (x_i - \mu)^2 \qquad \text{(A5–2)}$$

if the data represent a population. For grouped data repesenting a sample, the mean and variance are

$$\bar{x} = \sum_i f_i x_i \qquad \text{(A5–3)}$$

and

$$s^2 = \frac{n}{n-1} \left[\sum_i f_i (x_i - \bar{x})^2 \right] . \qquad \text{(A5–4)}$$

Computationally easier formulas for the variance are

$$\sigma^2 = \sum_i f_i x_i^2 - \mu^2 \qquad \text{(A5–5)}$$

for a population and

$$s^2 = \frac{n}{n-1} \left[\sum_i f_i x_i^2 - \bar{x}^2 \right] \qquad \text{(A5–6)}$$

for a sample.

ISP COMMANDS

The mean and variance of grouped data can be found directly using the ISP command SEXPV. Given a column of x_i values and a corresponding column of relative frequencies f_i, the computer calculates the mean and variance.

In order to obtain the mean and variance indirectly by actually applying the formulas for grouped data, appropriate transformations can be used in ISP. For example, TMUL can be used to multiply the x_i values and the relative frequencies, and the sum of the resulting column is the mean. This sort of use of transformations is illustrated in the following example.

SOLVED EXAMPLE

Suppose we want to compute the mean and variance of the frequency distribution of statistics grades given in Table A5.1. The first thing we need to do is divide each absolute frequency by the total number of students, which is 62 (in ISP this can be done by the command TDCN, which divides all elements of a column by a constant). The result is given in Table A5.2. From the columns of x-values and relative frequencies, we can compute the mean and variance from the command SEXPV, as shown in Table A5.2.

Alternatively, we could use first (A5–1) and then either (A5–2) or (A5–5). The computations required to find the mean via (A5–1) can be seen in Table A5.3 (the ISP command TMUL, which multiplies two columns term by term, has been used). The mean is 2.387, the sum of the fx column. Note that this agrees with the result in Table A5.2.

In order to find the variance using formula (A5–2), we make the computations shown in Table A5.4 (the ISP command TCON creates column 3; TSUB creates column 4; TCC creates column 5; and TMUL creates column 6). The variance is 1.4652, the sum of the final column. Another possibility is to find the variance by using formula (A5–5). The computations required are given in Table A5.5. The variances from Tables A5.4 and A5.5 differ slightly from that in Table A5.2 because of rounding error.

TABLE A5.1   Frequency distribution of statistics grades

Grade	Number of students (Absolute frequency)
0	5
1	10
2	16
3	18
4	13
	62

TABLE A5.2   Computation of relative frequencies and the mean and variance of statistics grades

⟨n⟩ PRINT

Observ.		x	Absol.	f
*	1*	0.000	5.000	0.081
*	2*	1.000	10.000	0.161
*	3*	2.000	16.000	0.258
*	4*	3.000	18.000	0.290
*	5*	4.000	13.000	0.210
Sum		10.0000	62.0000	1.0000
n =		5	5	5
Mean		2.0000	12.4000	0.2000

⟨n⟩ SEXPV
***** Computes expected values *****
Enter column number for random variable X ? 1
Enter column number for probabilities ? 3
Mean =   2.3870968   Variance =   1.4630593

TABLE A5.3   Computation of mean of statistics grades

⟨n⟩ PRINT

Observ.		x	f	fx
*	1*	0.000	0.081	0.000
*	2*	1.000	0.161	0.161
*	3*	2.000	0.258	0.516
*	4*	3.000	0.290	0.870
*	5*	4.000	0.210	0.840
Sum		10.0000	1.0000	2.3870
n =		5	5	5
Mean		2.0000	0.2000	0.4774

TABLE A5.4   Computation of variance of statistics grades

⟨n⟩ PRINT

Observ.		x	f	μ	$x - \mu$	$(x - \mu)^2$	$f(x - \mu)^2$
*	1*	0.000	0.081	2.3870	-2.387	5.698	0.462
*	2*	1.000	0.161	2.3870	-1.387	1.924	0.310
*	3*	2.000	0.258	2.3870	-0.387	0.150	0.039
*	4*	3.000	0.290	2.3870	0.613	0.376	0.109
*	5*	4.000	0.210	2.3870	1.613	2.602	0.546
Sum		10.0000	1.0000	11.9350	-1.9350	10.7488	1.4652
n =		5	5	5	5	5	5
Mean		2.0000	0.2000	2.3870	-0.3870	2.1498	0.2930

TABLE A5.5   Alternate calculation of variance of statistics grades

Observ.		x	f	x^2	fx^2
*	1*	0.000	0.081	0.000	0.000
*	2*	1.000	0.161	1.000	0.161
*	3*	2.000	0.258	4.000	1.032
*	4*	3.000	0.290	9.000	2.610
*	5*	4.000	0.210	16.000	3.360
Sum		10.0000	1.0000	30.0000	7.1630
n =		5	5	5	5
Mean		2.0000	0.2000	6.0000	1.4325

$$\sigma^2 = \sum_i f_i x_i^2 - \mu^2$$

$$= 7.163 - 2.387^2 = 1.465 \quad .$$

EXERCISES

A5.1.

    In a survey of mobility, it is found that 42 families have never moved from their initial home, 57 families have moved once, 35 families have moved twice, 18 families have moved three times, 9 families have moved 4 times, and 2 families have moved five times. Enter the data into ISP, convert the absolute frequencies to relative frequencies, and use SEXPV to find the mean and variance.

A5.2.

    For the salary data in Table A1.7, use PFREQ to obtain a relative frequency distribution. Then, letting the midpoint of each class represent that class, enter these values (the midpoints) and the relative frequencies. Use SEXPV to find the mean and variance for the grouped data. How do these values compare with the mean and variance found directly from the raw data without grouping?

A5.3

    For the data in Table A1.6, construct a relative frequency distribution by using PFREQ. Enter the values and relative frequencies appropriately and use SEXPV to find the mean and variance. Compare these values with the mean and variance found directly from the raw data without grouping. Finally, does this last comparison yield the same type of result found in the similar comparison in Exercise A5.2?

A6
The Empirical Rule and Chebyshev's Theorem

Suppose that we do not know the frequency distribution of a data set, but we do have some idea of the shape of the distribution. If it is roughly bell-shaped, then approximately 68 percent of the data are within one standard deviation of the mean; approximately 95 percent of the data are within two standard deviations of the mean; and approximately 99.7 percent, or almost all of the data, are within three standard deviations of the mean. These percentages are based on what is called "the empirical rule," which allows us to obtain approximate figures for distributions that are roughly bell-shaped.

If the shape of the distribution is not known, a mathematical result known as Chebyshev's theorem provides some information. This theorem tells us that the relative frequency of data between $\mu - k\sigma$ and $\mu + k\sigma$ (that is, from k standard deviations below the mean to k standard deviations above the mean) is at least (not exactly, but at least)

$$1 - \frac{1}{k^2} \ .$$

ISP COMMANDS

Since only simple arithmetic manipulations are involved, no special ISP commands are needed. The mean and standard deviation are used in applying the empirical rule or Chebyshev's theorem, and these summary measures can be found by using the commands SMEAN and SVAR, as discussed in Chapters A3 and A4. Later we will introduce commands to deal with a generalization of our simple formulas for the empirical rule and to illustrate Chebyshev's theorem and the empirical rule through the use of simulation.

SOLVED EXAMPLE

Suppose that the mean IQ score in a large population of students is 110 and the standard deviation of IQ scores in this population is 10. Approximately what percentage of the students have IQ scores between 90 and 130? Distributions of IQ scores are known to be roughly bell-shaped, so that the empirical rule can be used. The lower value, 90, is two standard deviations below the mean of 110, and 130 is two standard deviations above the mean of 110. Thus, according to the empirical rule, approximately 95 percent of the students have IQ scores between 90 and 130. Incidentally, the remaining 5 percent should be roughly evenly divided, with 2.5 percent of the students having IQ scores below 90 and 2.5 percent having IQ scores above 130.

If nothing is known about the shape of the distribution, Chebyshev's theorem can be used even though the empirical rule cannot. This would tell us that at least

$$1 - \frac{1}{2^2} = 0.75 \quad ,$$

or 75 percent of the IQ scores, would be between 90 and 130. This is a much weaker statement than the one we are able to make from the empirical rule. The empirical rule provides a stronger statement because it uses a stronger assumption (a bell-shaped frequency distribution).

EXERCISES

A6.1.

Suppose that the mean family income in a community is $25,000 and the standard deviation is $10,000. What can you say about the percentage of families with incomes between $5000 and $45,000 if the distribution of incomes is roughly bell-shaped? What can you say if the distribution is definitely *not* bell-shaped?

A6.2.

For the data in Table A1.6, find the mean and variance and then determine, by counting, the exact relative frequency within two standard deviations of the mean. How does this value compare with the estimate obtained from the empirical rule and with the bound provided by Chebyshev's theorem?

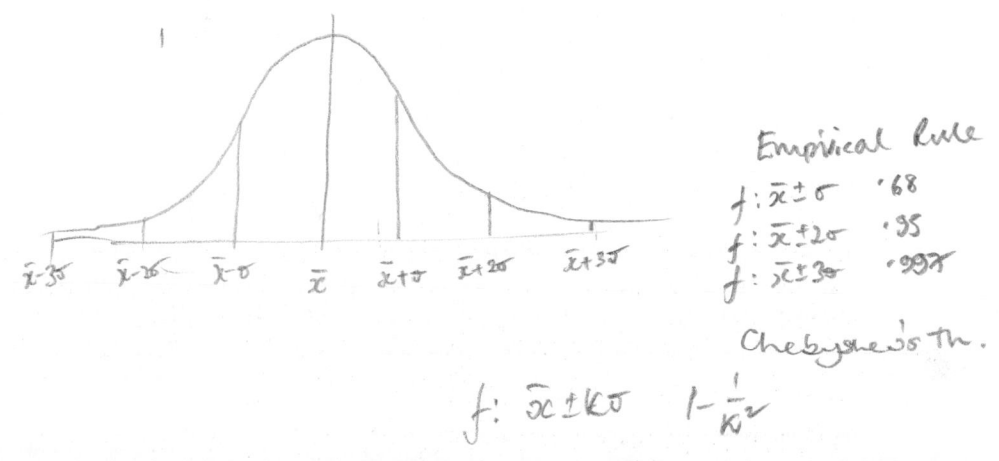

A7
Standardizing Data

A set of data can be standardized by subtracting the mean from each data value and then dividing by the standard deviation. This process of standardizing data is also called coding, and the resulting standardized values are often called z-scores, or z-values. The standardized value corresponding to a data point x is

$$z = \frac{x - \mu}{\sigma} \tag{A7-1}$$

if we are dealing with a population and

$$z = \frac{x - \bar{x}}{s} \tag{A7-2}$$

if we are dealing with a sample.

    The mean of a set of standardized values is always zero and the standard deviation is always one. In essence, we change the scale so that the data fluctuate around 0 with a standard deviation of 1. Then we know that a positive z-value indicates a data point above the mean, a negative z-value indicates a data point below the mean, and a z-value of, say, -3 or $+3$ is quite extreme because it indicates a data point 3 standard deviations below or above the mean. A standardized value tells us exactly how many standard deviations above of below the mean the data point is.

ISP COMMANDS

The command TSTD provides standardized values for all elements of a data set (a column of data in ISP). To go through the individual steps, first use SMEAN and SVAR to find the mean and standard deviation. Then use TSCN to subtract a constant (the mean) from each data point and use TDCN to divide the resulting values by another constant (the standard deviation).

SOLVED EXAMPLE

Suppose we want to standardize the temperature data in Table A7.1. In both cities, the mean temperature is 70, but the standard deviation is much larger for Anchorage than for St. Thomas. Standardized values for the temperatures in St. Thomas and Anchorage can be found directly by using the command TSTD. The resulting standardized values are given in Table A7.2. Note, for example, that a temperature of 66 in St. Thomas (the second observation in St. Thomas) corresponds to $z = -1.414$, whereas a temperature of 66 in Anchorage (the last observation in Anchorage) corresponds to $z = -0.141$. In St. Thomas, 66 is 1.414 standard deviations below the mean for our little sample of 6 temperatures. In Anchorage, it is only 0.141 standard deviations below the mean.

Alternatively, the standardized values can be found by using transformations. For St. Thomas, we need to subtract the mean of 70 from each temperature reading and then divide the difference by the standard deviation of 2.828. The result can be seen in the last column of Table A7.3 (the commands TSCN and TDCN were used). In a similar way, using the Anchorage mean temperature of 70 and standard deviation of 28.36, the standardized values of the temperature readings in Anchorage can be found, as can be seen in Table A7.4.

TABLE A7.1   Temperatures in St. Thomas and Anchorage

Temperatures in St. Thomas	Temperatures in Anchorage
69	47
66	42
68	55
73	108
73	102
71	66

TABLE A7.2   Standardized values for temperatures

⟨n⟩ PRINT

Observ.		St. Thomas	Anchorage
*	1*	−0.354	−0.811
*	2*	−1.414	−0.987
*	3*	−0.707	−0.529
*	4*	1.061	1.340
*	5*	1.061	1.128
*	6*	0.354	−0.141
Sum		0.0000	0.0000
n =		6	6
Mean		0.0000	0.0000

TABLE A7.3   Standardized temperatures for St. Thomas

⟨n⟩ PRINT

Observ.	x	x−x̄	z
* 1*	69.000	−1.000	−0.354
* 2*	66.000	−4.000	−1.414
* 3*	68.000	−2.000	−0.707
* 4*	73.000	3.000	1.061
* 5*	73.000	3.000	1.061
* 6*	71.000	1.000	0.354
Sum	420.0000	0.0000	0.0000
n =	6	6	6
Mean	70.0000	0.0000	0.0000

TABLE A7.4   Standardized temperatures for Anchorage

⟨n⟩ PRINT

Observ.	x	x−x̄	z
* 1*	47.000	−23.000	−0.811
* 2*	42.000	−28.000	−0.987
* 3*	55.000	−15.000	−0.529
* 4*	108.000	38.000	1.340
* 5*	102.000	32.000	1.128
* 6*	66.000	−4.000	−0.141
Sum	420.0000	0.0000	0.0000
n =	6	6	6
Mean	70.0000	0.0000	0.0000

EXERCISES

A7.1.

Convert the accident data in Table A1.5 to standardized values. If there are 5 accidents in a week, how many standard deviations above or below the mean is this?

A7.2.

Use TSTD to standardize the ages given in Exercise A4.1. Then use SMEAN and SVAR on the standardized values to verify that their mean is zero and their variance is one.

A7.3.

For the salary data in Table A1.7, how many standard deviations above or below the mean is the first salary, $29,400?

A7.4.

Two friends take two different exams. The first gets a score of 80, and the second scores 85. However, in terms of standardized values, the z-value for the first is 1.5 and the z-value for the second is -1. What can be said about the performance of the two friends?

Summary of
Part A

The easiest way to make sense out of large sets of data is to summarize them in some fashion. Such summarization of data is the role of descriptive statistics. Many different ways are available to describe and summarize data, graphically or otherwise. It is impossible to cover all available methods here, and the purpose of Part A has been to discuss the types of descriptive methods that traditionally have been used most often in statistics.

Frequency distributions and their graphical counterpart, histograms, are convenient ways to help us comprehend the information contained in a set of data. They convey the "overall picture" of the data. Often we are also interested in particular aspects of the data, in which case summary measures may be helpful. Measures of location tell us something about the "center" of a set of data or a "typical member" of the set. Measures of dispersion tell us how spread out the data are. Yet other measures tell us about other aspects of the data, such as skewness. Two measures of special interest are the mean and standard deviation (or its square, the variance). The empirical rule and Chebyshev's theorem involve the mean and standard deviation, and these two measures are essential in standardizing data. Descriptive measures such as the mean and standard deviation are also useful in inferential statistics, as you shall see later in this book.

VOCABULARY LIST

absolute frequency

average

bell-shaped distribution

Chebyshev's theorem

class

class interval

coding

coefficient of kurtosis

coefficient of skewness

coefficient of variation

cumulative frequency

cumulative histogram

cumulative relative frequency

data

descriptive statistics

deviation from the mean

dispersion

empirical rule

frequency

frequency distribution

grouped data

histogram

kurtosis

location

mean

measure of dispersion

measure of location

median

midpoint

mode

ordered data

parameter

pie chart

population

population mean

population standard deviation

population variance

range

ranked data

relative frequency

sample

sample mean

sample standard deviation

sample variance

skewed distribution

skewness

standard deviation

standardized values

standardizing data

statistic

statistics

summary statistic

symmetric distribution

variance

z-scores (z-values)

LIST OF SYMBOLS

x_i – the ith number in a set of numbers (data set)

Σ – summation

N – population size (total number of units in the population)

μ – the mean of the population

σ^2 – the variance of the population

σ – the standard deviation of the population

n – sample size (total number of units in the sample)

$\bar{x}$ – the sample mean

s^2 – the sample variance

s – the sample standard deviation

f_i – the relative frequency of class i or value i

z – a standardized value

FORMULAS

$$\mu = \frac{\sum_{i=1}^{N} x_i}{n}$$

$$\mu = \sum_{i} f_i x_i \qquad \text{(for grouped data)}$$

relative frequencies

$$\sigma^2 = \frac{\sum_{i=1}^{N} (x_i - \mu)^2}{N}$$

$$\sigma^2 = \frac{N \sum_{i=1}^{N} x_i^2 - \left(\sum_{i=1}^{N} x_i\right)^2}{N^2}$$

$$\sigma^2 = \sum_{i} f_i (x_i - \mu)^2 \qquad \text{(for grouped data)}$$

$$\sigma^2 = \sum_{i} f_i x_i^2 - \mu^2 \qquad \text{(for grouped data)}$$

$$\sigma = \sqrt{\frac{\sum_{i=1}^{N} (x_i - \mu)^2}{N}}$$

$$\sigma = \sqrt{\frac{N \sum_{i=1}^{N} x_i^2 - \left(\sum_{i=1}^{N} x_i\right)^2}{N^2}}$$

$$\sigma = \sqrt{\sum_{i} f_i (x_i - \mu)^2} \qquad \text{(for grouped data)}$$

$$\sigma = \sqrt{\sum_{i} f_i x_i^2 - \mu^2} \qquad \text{(for grouped data)}$$

$$\bar{x} \;\; = \;\; \frac{\displaystyle\sum_{i=1}^{n} x_i}{n}$$

$$\bar{x} \;\; = \;\; \sum_i f_i x_i \qquad\qquad\qquad\qquad \text{(for grouped data)}$$

$$s^2 \;\; = \;\; \frac{\displaystyle\sum_{i=1}^{n} (x_i - \bar{x})^2}{n-1}$$

$$s^2 \;\; = \;\; \frac{n \displaystyle\sum_{i=1}^{n} x_i^2 - \left(\sum_{i=1}^{n} x_i\right)^2}{n(n-1)}$$

$$s^2 \;\; = \;\; \frac{n}{n-1}\left[\sum_i f_i(x_i - \bar{x})^2\right] \qquad\qquad \text{(for grouped data)}$$

$$s^2 \;\; = \;\; \frac{n}{n-1}\left[\sum_i f_i x_i^2 - \bar{x}^2\right] \qquad\qquad \text{(for grouped data)}$$

$$s \;\; = \;\; \sqrt{\frac{\displaystyle\sum_{i=1}^{n} (x_i - \bar{x})^2}{n-1}}$$

$$s \;\; = \;\; \sqrt{\frac{n \displaystyle\sum_{i=1}^{n} x_i^2 - \left(\sum_{i=1}^{n} x_i\right)^2}{n(n-1)}}$$

$$s \;\; = \;\; \sqrt{\frac{n}{n-1}\left[\sum_i f_i(x_i - \bar{x})^2\right]} \qquad\qquad \text{(for grouped data}$$

$$s \;\; = \;\; \sqrt{\frac{n}{n-1}\left[\sum_i f_i x_i^2 - \bar{x}^2\right]} \qquad\qquad \text{(for grouped data)}$$

$$\text{Coefficient of Variation} = \frac{\sigma}{\mu} \qquad \text{(for a population)}$$

$$\text{Coefficient of Variation} = \frac{s}{\bar{x}} \qquad \text{(for a sample)}$$

$$\text{Coefficient of Skewness} = \frac{\sum_{i=1}^{n}(x_i - \mu)^3}{N\sigma^3} \qquad \text{(for a population)}$$

Is the distⁿ Skewed or Symmetrical.

$$\text{Coefficient of Skewness} = \frac{\sum_{i=1}^{n}(x_i - \bar{x})^3}{ns^3} \qquad \text{(for a sample)}$$

$$\text{Coefficient of Kurtosis} = \frac{\sum_{i=1}^{n}(x_i - \mu)^4}{N\sigma^4} \qquad \text{(for a population)}$$

Is

$$\text{Coefficient of Kurtosis} = \frac{\sum_{i=1}^{n}(x_i - \bar{x})^4}{ns^4} \qquad \text{(for a sample)}$$

How fat or thin are the tails

$$z = \frac{x - \mu}{\sigma} \qquad \text{(for a population)}$$

$$z = \frac{x - \bar{x}}{s} \qquad \text{(for a sample)}$$

Coded data standardized

RELEVANT ISP COMMANDS

For descriptive statistics, commands from the P, S, C, and T groups of commands are useful. Included in the P group are PFREQ, which provides a frequency distribution, and PHIST, which constructs a histogram. The PRINT command can be used to print a data set, PDIST enables you to compare a frequency distribution to various theoretical distributions that will be discussed later in this book, and PROBD gives graphs of the theoretical distributions.

The easiest way to compute summary statistics in ISP is through the following commands from the S group.

SMEAN: Computes the mean
SMEDI: Computes the median
SMODE: Computes the mode
SRANG: Computes the range
SVAR: Computes the variance and standard deviation
SUMST: Computes major summary statistics
SKKU: Computes the skewness and kurtosis
SEXPV: Computes expected values

SMEAN, SMEDI, and SMODE provide three measures of location. Measures of dispersion are provided by SRANG and SVAR, and some of the major summary statistics are given by a single command, SUMST. The coefficients of skewness and kurtosis are computed by SKKU. Finally, to calculate the mean and variance of grouped data, use SEXPV.

Of course, it is possible to carry out computations of summary statistics and to generate frequency distributions without using the shortcut provided by the P and S commands. Doing so can provide a better understanding of the various measures encountered in descriptive statistics. The computing commands (C commands) allow you to make any arithmetic computations you may wish.

CADD: Adds n numbers
CALC: Hand calculator
CDIV: Divides two numbers
CEXP: Gives e raised to any power
CFAC: Factorials
CLOG: Natural and common logarithms
CMULT: Multiplies two numbers
CPOW: Raises a number to a power
CSQQ: Gives the square of a number
CSQRT: Gives the square root of a number
CSUB: Subtracts two numbers
CSXX: Sums n squared numbers

The T commands are also quite helpful because they enable you to perform computations on entire data sets (columns of numbers) at once. These transformations are particularly useful in the computation of the mean, the variance, and other summary statistics. Here is a list of the transformations available in ISP.

TACN:	Adds a constant to all elements of a column
TADD:	Adds two columns term by term
TCC:	Gives the square of all elements of a column
TCON:	Sets a column to a constant
TDCN:	Divides all elements of a column by a constant
TDIF:	Gives the first and second differences of elements in a column
TDIV:	Divides two columns term by term
TEXP:	Gives e raised to the power of each element of a column
TLOG:	Gives the logarithm of all elements of a column
TMCN:	Multiplies all elements of a column by a constant
TMUL:	Multiplies two columns term by term
TPOW:	Raises a column to a stated power
TSCN:	Subtracts a constant from all elements of a column
TSORC:	Puts the elements of a column in numerical order
TSORM:	Rearranges the elements of all columns in numerical order
TSQT:	Gives the square root of all elements of a column
TSTD:	Gives standardized values for all elements of a column
TSUB:	Subtracts two columns term by term

The MD commands in ISP provide brief summaries of the concepts of descriptive statistics. The seven MD modules correspond to the seven chapters of Part A.

MDA1:	Frequency distributions
MDA2:	Histograms and graphs
MDA3:	Measures of location
MDA4:	Measures of dispersion
MDA5:	Grouped data
MDA6:	The empirical rule and Chebyshev's theorem
MDA7:	Standardizing data

PART B

Probability and Probability Distributions

In Part A several methods were presented to describe or summarize data. These are referred to as methods of descriptive statistics. While descriptive methods are important, we are often concerned with inferential statistics, which consists of methods for using information from a sample to make inferences about the population from which the sample was drawn. For example, a typical inferential problem would be to estimate the average income in a community of 30,000 people from a sample of incomes of 500 randomly selected inhabitants. The mean of the sample of the 500 inhabitants is a way of describing or summarizing the data from the sample. However, in addition to simply describing or summarizing the data, we may attempt to use the sample data to make an inference about the average income of all 30,000 inhabitants.

When we make inferences, we cannot be absolutely certain of being correct. Similarly, when making decisions in an uncertain world we cannot be sure that our decisions will turn out after the fact to be the best possible choices. In order to make better, more intelligent inferences and decisions, we have to be able to measure and deal with uncertainty. This is why we are going to discuss the concept of probability in Part B. Probability is the language of uncertainty. In this respect Part B provides a transition from descriptive statistics (Part A) to inferential statistics (Parts C, D, E, and F). Probability can be used to quantify and measure uncertainty in a way that is useful for improving decision making.

Chapters B1, B2, B3, and B4 discuss probability and some basic probability rules. Chapter B5 introduces probability distributions, while Chapter B6 examines how expected values (in particular, the mean and variance of probability distributions) can be found. Finally, three of the most commonly encountered theoretical probability distributions (binomial, Poisson, and normal) are presented in Chapters B7, B8, and B9, respectively.

B1
Introduction to Probability

Most people understand and use words such as "probably," "likely," and "chances" but may be intimidated by the notion of probability. In this chapter we concentrate on defining probability using three different interpretations that cover the essence of the way probability is used in practice. In the classical interpretation, probabilities are based on the idea of equally likely outcomes. This interpretation is applicable in many games of chance (for example, roulette), lotteries, and random sampling (which will be discussed in a later chapter). In the relative frequency interpretation, probability is defined as the proportion of times an event occurs in repeated trials:

$$\text{Probability of an event} = \frac{\begin{array}{c}\text{Number of times the event happens}\\ \text{in a long series of repeated trials}\end{array}}{\text{Number of trials}} \quad . \quad \text{(B1–1)}$$

The larger the number of trials, the more accurate the estimated probability. This is intuitively reasonable and can be formalized in terms of the law of large numbers, which states that as the number of trials increases, the relative frequency of times an event occurs tends to get closer and closer to the probability of that event. For example, actuarial tables that estimate the probability of dying within the next year are based on extensive data. The third interpretation of probability is subjective in nature, estimating probabilities on the basis of your own judgments about how likely certain events are.

Sometimes we express uncertainty in terms of odds instead of probability. The conversion from probability to odds is a simple calculation:

$$\text{Odds in favor of an event} = \frac{\text{Probability of the event}}{1 - \text{Probability of the event}} \quad . \quad \text{(B1–2)}$$

We can also find the probability corresponding to a value for odds:

$$\text{Probability of an event} = \frac{\text{Odds in favor of the event}}{1 + \text{Odds in favor of the event}} \quad . \quad \text{(B1–3)}$$

ISP COMMANDS

ISP is not needed for the simple computations involved in the three interpretations of probability, although it can be used for such computations. The primary role of ISP at this point is to provide you with a feel for uncertainty, probability, and the speed with which a relative frequency approaches the probability of an event. This can be accomplished by experimenting with several games in ISP. The following ISP commands are suggested:

GCOIN:	Simulates the tossing of up to 20 coins.
GDIE:	Simulates the throwing of up to 20 dice.
GMARB:	Simulates the sampling of marbles from an urn. The number of colors (up to four) or marbles can be specified by the user. Sampling marbles can be done with or without replacement.
GTACK:	Simulates the tossing of a thumbtack. The probability of "point up" and "point down" is determined randomly (within a given range) by the program. The purpose of using the program is to estimate this probability experimentally.
GROUL:	Simulates the game of roulette.
GLOTO:	Simulates the game of LOTTO (the most popular lottery game in France).
GCARD:	Simulates playing with a deck of cards.

These routines help you to better understand probabilities and the law of large numbers. They can be used throughout Part B.

SOLVED EXAMPLE

Suppose that we throw a thumbtack 100 times in order to estimate the probability that it will land "point up." Figure B1.1 shows 100 throws of a thumbtack simulated by the computer via GTACK. The estimated probability of "point up" is 44/100 = 0.44. Thus,

FIGURE B1.1   Some output from the use of GTACK to simulate the tossing of a thumbtack 100 times.

Toss			
1	POINT UP	7	POINT UP
2	POINT DOWN	8	POINT UP
3	POINT DOWN	9	POINT DOWN
4	POINT UP	10	POINT UP
5	POINT DOWN	11	POINT UP
6	POINT DOWN	12	POINT UP

Continued

FIGURE B1.1 **Some output from the use of GTACK to simulate the tossing of a thumbtack 100 times. (cont.)**

13	POINT UP	57	POINT UP
14	POINT UP	58	POINT DOWN
15	POINT DOWN	59	POINT DOWN
16	POINT UP	60	POINT DOWN
17	POINT DOWN	61	POINT DOWN
18	POINT UP	62	POINT UP
19	POINT DOWN	63	POINT DOWN
20	POINT UP	64	POINT DOWN
21	POINT DOWN	65	POINT UP
22	POINT UP	66	POINT UP
23	POINT UP	67	POINT DOWN
24	POINT DOWN	68	POINT DOWN
25	POINT UP	69	POINT UP
26	POINT DOWN	70	POINT DOWN
27	POINT DOWN	71	POINT DOWN
28	POINT DOWN	72	POINT UP
29	POINT UP	73	POINT DOWN
30	POINT UP	74	POINT DOWN
31	POINT DOWN	75	POINT DOWN
32	POINT DOWN	76	POINT DOWN
33	POINT DOWN	77	POINT DOWN
34	POINT UP	78	POINT DOWN
35	POINT DOWN	79	POINT UP
36	POINT UP	80	POINT DOWN
37	POINT UP	81	POINT UP
38	POINT UP	82	POINT DOWN
39	POINT UP	83	POINT UP
40	POINT UP	84	POINT DOWN
41	POINT DOWN	85	POINT DOWN
42	POINT UP	86	POINT DOWN
43	POINT UP	87	POINT DOWN
44	POINT DOWN	88	POINT DOWN
45	POINT UP	89	POINT DOWN
46	POINT UP	90	POINT DOWN
47	POINT DOWN	91	POINT DOWN
48	POINT DOWN	92	POINT UP
49	POINT UP	93	POINT DOWN
50	POINT DOWN	94	POINT DOWN
51	POINT UP	95	POINT DOWN
52	POINT UP	96	POINT UP
53	POINT UP	97	POINT DOWN
54	POINT DOWN	98	POINT DOWN
55	POINT DOWN	99	POINT DOWN
56	POINT UP	100	POINT UP

Number of POINT UP	Absolute Freq.	Relative Freq.
0	56	.560
1	44	.440
	100	1.000

''point down'' is more likely than ''point up,'' and the estimated odds in favor of ''point down'' are (0.56/0.44)-to-1, or about 1.27-to-1.

Of course, if we repeated this experiment we might get a slightly different result. In this case, it turns out that if we were to continue to simulate throws, in the long run about 43.6% of the throws would result in the thumbtack landing ''point up.''

EXERCISES

B1.1.

If a die is fair (that is, each of the six sides is equally likely to come up when the die is tossed), what is the probability that a 4 will come up if the die is tossed? Answer this question by using the classical interpretation of probability. Then use GDIE to simulate 100 tosses of a single die and estimate the probability of a 4 coming up by using the relative frequency interpretation of probability. Compare the two values you have obtained and explain any difference between them.

B1.2.

Use GCOIN to simulate the tossing of 10 coins, and toss the 10 coins 50 times. From the results, determine an estimate of the probability that exactly 5 of the 10 coins will come up heads. Why is this only an estimate, and not exactly the ''true'' probability? Toss the coins (via GCOIN) 50 more times and determine a new estimate based on all 100 tosses of the 10 coins.

B1.3.

For a person of a particular age and sex, actuarial tables indicate that the probability of living at least another year is 0.98. From this actuarial value, what are the odds in favor of living at least another year? If the person in question has a physical examination and the doctor thinks that her subjective probability that the person will live at least another year is 0.80, what are the odds corresponding to this subjective judgment?

B2
Rules of Probability

Probability is a well-established mathematical discipline. The purpose of this and the following chapters is to describe some basic rules of probability that will be useful in inferential statistics. Before we describe the rules, though, you should know the following terms.

1. Experiment: Any process used for obtaining a measurement or observation.

2. Outcome: A specific result of an experiment.

3. Event: A collection of outcomes of an experiment.

4. Sample Space: All possible outcomes of an experiment.

   Now we are prepared to present some basic rules of probability, using P to denote probability. Thus, $P(A)$ represents the probability that event A occurs, $P(B)$ the probability that B occurs, and so on.

1. Probabilities cannot be negative. For any event A,

$$P(A) \geq 0 \quad . \tag{B2-1}$$

2. Probabilities cannot be larger than one. For any event A,

$$P(A) \leq 1 \quad . \tag{B2-2}$$

3. *Addition Rule*. If two events A and B are mutually exclusive (that is, they cannot *both* occur), then the probability that either A or B occurs is

$$P(A \text{ or } B) = P(A) + P(B) \quad . \tag{B2-3}$$

4. Modified addition rule for non-mutually exclusive events:

$$P(A \text{ or } B) = P(A) + P(B) - P(A \text{ and } B) \quad , \tag{B2-4}$$

where $P(A$ and $B)$ is the probability that both A and B occur.

5. The probability of a complement (that is, that A does not occur):

$$P(\text{not } A) = 1 - P(A) \quad . \tag{B2–5}$$

6. *Multiplication Rule*. If two events A and B are independent (that is, whether or not A occurs has no bearing on whether or not B occurs, and vice versa), then the probability that both A and B occur is

$$P(A \text{ and } B) = P(A)\, P(B) \quad . \tag{B2–6}$$

The probability $P(A$ and $B)$ is called the joint probability of A and B. The independence requirement is very important. If the events are not independent, we cannot find their joint probability by multiplying their individual probabilities. For non-independent events, we need to consider conditional probabilities, which will be discussed in the next chapter.

ISP COMMANDS

The computing commands of ISP (the C group) are relevant here. In addition, commands such as GCOIN, GDIE, GMARB, GTACK, GROUL, GLOTO, and GCARD can be used to gain experience with uncertainty and to illustrate some of the concepts discussed here.

SOLVED EXAMPLE

Suppose that three candidates are running in an election and that 50% of the voters prefer Candidate I, 30% prefer Candidate II, and 20% prefer Candidate III. (No voters are indifferent in this population.) If a voter is chosen at random (random selection will be defined formally in a later chapter), the probability is 0.50 that the voter prefers I, 0.30 that the voter prefers II, and 0.20 that the voter prefers III. Preferring I and preferring II are mutually exclusive, so we have

$$P(\text{prefers I or II}) = P(\text{prefers I}) + P(\text{prefers II})$$
$$= 0.50 + 0.30 = 0.80 \quad .$$

Now suppose that a second voter is selected independently of the first. The probability that both voters prefer II is

$$P(\text{both prefer II}) = P(\text{first prefers II})\, P(\text{second prefers II})$$
$$= (0.30)\,(0.30) = 0.09 \quad .$$

What is the probability that one voter prefers II and the other voter prefers I? This can happen in two mutually exclusive ways: either the first voter prefers II and the second prefers I, or the first prefers I and the second prefers II. The probabilities for these two possibilities are

$$P(\text{first prefers II and second prefers I}) = P(\text{first prefers II}) \, P(\text{second prefers I})$$
$$= (0.30) \, (0.50) = 0.15$$

and

$$P(\text{first prefers I and second prefers II}) = P(\text{first prefers I}) \, P(\text{second prefers II})$$
$$= (0.50) \, (0.30) = 0.15 \quad .$$

Thus,

$$P(\text{one prefers II, one prefers I}) = 0.15 + 0.15 = 0.30 \quad .$$

To calculate this last probability, we used the multiplication rule to get 0.15 and 0.15 and the addition rule to find the final answer, 0.30.

EXERCISES

B2.1.

Use GCOIN to simulate the tossing of 5 coins, and toss the 5 coins 50 times. From the results, determine an estimate of the probability that 2 or fewer of the 5 coins will come up heads.

B2.2.

Use the ISP command GTACK to simulate the tossing of a thumbtack 100 times. From the results, estimate the probability that the tack lands "point up." What does this tell you about the probability that the tack lands "point down?" What about the probability that when the tack is tossed twice, it lands "point up" both times?

B2.3.

Of the radios produced at a certain factory, 5 percent have defective speakers and 7 percent have defective tuning. Moreover, 2 percent have *both* defective speakers and defective tuning. What is the probability that a radio produced at this factory has at least one of these two defects?

B3

Conditional Probability

The concept of independence is very important in statistics, and the multiplication rule presented in Chapter B2 requires independence. However, many events in the real world are related in some fashion to each other and are therefore not independent. A modified multiplication rule for dependent events is

$$P(A \text{ and } B) = P(A) \, P(B|A) \quad , \tag{B3–1}$$

where $P(B|A)$ represents the probability of occurrence of B given that A has already occurred. That is, it is the probability of B *conditional* upon the occurrence of A. Thus, the probability $P(B|A)$ is called a conditional probability.

By manipulating $(B3–1)$, we can arrive at a formula for a conditional probability:

$$P(B|A) = \frac{P(A \text{ and } B)}{P(A)} \quad , \tag{B3–2}$$

where $P(A \text{ and } B)$ is the joint probability of A and B (that is, the probability that both A and B occur) and $P(A)$ is simply the probability that A occurs. Notice that the multiplication rule from the preceding chapter is a special case of (B3–1). If A and B are independent, then A does not influence the probability of B, and $P(B|A) = P(B)$. In fact, independence is usually defined in this fashion. The events A and B are said to be independent if

$$P(B|A) = P(B) \quad . \tag{B3–3}$$

ISP COMMANDS

As in Chapter B2, the computing commands of ISP are relevant here.

SOLVED EXAMPLE

Suppose that the cars produced at a certain plant sometimes have defective brakes, defective steering, or both. We know that the probability that a car has defective brakes is 0.10. If a car has defective brakes, the probability is 0.40 that it also has defective steering. However, if the brakes are not defective, the probability of defective steering is only 0.10. We want to find the joint probabilities for the four possible combinations: defective brakes and defective steering, defective brakes and good steering, good brakes and defective steering, and good brakes and good steering.

The probabilities that have been given are

$$P(\text{defective brakes}) = 0.10,$$
$$P(\text{defective steering}|\text{defective brakes}) = 0.40,$$
$$\text{and } P(\text{defective steering}|\text{good brakes}) = 0.10.$$

Therefore, from (B3–1),

$$P(\text{defective brakes and defective steering})$$
$$= P(\text{defective brakes}) \, P(\text{defective steering}|\text{defective brakes})$$
$$= 0.10 \, (0.40) = 0.04.$$

But since $P(\text{defective steering}|\text{defective brakes}) = 0.40$, we must have

$$P(\text{good steering}|\text{defective brakes}) = 1 - 0.40 = 0.60.$$

Thus, using (B3–1) once again,

$$P(\text{defective brakes and good steering})$$
$$= P(\text{defective brakes}) \, P(\text{good steering}|\text{defective brakes})$$
$$= 0.10 \, (0.60) = 0.06.$$

What about the events involving good brakes? The probability of good brakes is

$$P(\text{good brakes}) = 1 - P(\text{defective brakes}) = 1 - 0.10 = 0.90.$$

Now,

$$P(\text{good brakes and defective steering})$$
$$= P(\text{good brakes}) \, P(\text{defective steering}|\text{good brakes})$$
$$= 0.90 \, (0.10) = 0.09.$$

Also,

$$P(\text{good steering}|\text{good brakes})$$
$$= 1 - P(\text{defective steering}|\text{good brakes})$$
$$= 1 - 0.10 = 0.90.$$

Therefore, the probability that both the brakes and the steering are operating properly is

$$P(\text{good brakes and good steering})$$
$$= P(\text{good brakes}) \, P(\text{good steering}|\text{good brakes})$$
$$= 0.90 \, (0.90) = 0.81.$$

It is convenient to summarize the joint probabilities in a table such as Table B3.1. The four joint probabilities are given in the body of the table. By adding across each row and down each column we can find the individual probabilities, which are also called marginal probabilities (note that they are on the margins of the table). The four marginal probabilities are

$$P(\text{defective brakes}) = 0.10,$$
$$P(\text{good brakes}) = 0.90,$$
$$P(\text{defective steering}) = 0.13,$$
$$\text{and } P(\text{good steering}) = 0.87.$$

TABLE B3.1 Joint probabilities for a car's brakes and steering

	Defective steering	Good steering	
Defective brakes	0.04	0.06	0.10
Good brakes	0.09	0.81	0.90
	0.13	0.87	

EXERCISES

B3.1.

Suppose that the probability of snow on any single day in March in Chicago is 0.05 and the probability of two consecutive days with snow is 0.02. If it snows on a March day, what is the probability that it will also snow on the following day?

B3.2.

Use GCOIN to toss a single coin 100 times. Count the number of times heads occurs and the number of times an occurrence of heads is followed by another occurrence of heads. Use this information to estimate the probability of getting heads given that the previous toss resulted in heads. Does it seem that successive tosses of the coin are independent?

B3.3.

In trading involving silver and gold, suppose that the probability that both metals increase in price on a given day is 0.40, the probability that neither increases in price is 0.30, the probability that gold increases and silver does not is 0.20, and the probability that silver increases and gold does not is 0.10. If silver increases in price, what is the probability that gold also increases in price? If gold increases in price, what is the probability that silver also increases in price?

B4
Bayes' Theorem

New information often causes us to alter our judgments about how likely an event is to occur. A convenient formula for the revision of probabilities is Bayes' theorem, which can be expressed as follows:

$$P(B|A) = \frac{P(B)P(A|B)}{P(B)P(A|B) + [1-P(B)] \, P(A|\text{not } B)} \, . \tag{B4-1}$$

The probability $P(B)$ is called a prior probability because it is the probability of B before, or prior to, seeing the new information. After, or posterior to, seeing the new information, $P(B|A)$ is the posterior probability of B. The other probabilities in (B4-1), $P(A|B)$ and $P(A|\text{not } B)$, are called the likelihoods. They represent the likelihood that the new information A would occur given B and the likelihood given "not B." The likelihoods tell us how useful the new information A is in helping us determine whether B will or will not occur.

Bayes' theorem can also be used to revise probabilities for a number of events B_1, B_2, ..., B_k:

$$P(B_i|A) = \frac{P(B_i)P(A|B_i)}{\sum_{j=1}^{k} P(B_j)P(A|B_j)} \, . \tag{B4-2}$$

The events B_1, ..., B_k must be mutually exclusive, and they must exhaust all possibilities (that is, exactly one of the k events will occur). This formula can be summarized in a table, as shown in Table B4.1.

ISP COMMANDS

The ISP command SBAYE can be used to revise probabilities with Bayes' theorem. If a single event is of interest, SBAYE asks for a prior probability $[P(B)]$ and two likelihoods $[P(A|B), P(A|\text{not } B)]$ and provides the posterior probability $[P(B|A)]$. With a number of

TABLE B4.1   Bayes' theorem in tabular form for a number of events $B_1, ..., B_k$

Event	Prior probability	Likelihood	Prior probability × likelihood	Posterior probability
B_1	$P(B_1)$	$P(A|B_1)$	$P(B_1)\,P(A|B_1)$	$P(B_1|A) = P(B_1)\,P(A|B_1)/\text{sum}$
B_2	$P(B_2)$	$P(A|B_2)$	$P(B_2)\,P(A|B_2)$	$P(B_2|A) = P(B_2)\,P(A|B_2)/\text{sum}$
.	.	.	.	.
.	.	.	.	.
.	.	.	.	.
B_k	$P(B_k)$	$P(A|B_k)$	$P(B_k)\,P(A|B_k)$	$P(B_k|A) = P(B_k)\,P(A|B_k)/\text{sum}$

$$\text{sum} = \sum_{j=1}^{k} P(B_j)P(A|B_j)$$

events, SBAYE asks for a number of prior probabilities and likelihoods and displays the tabular form of Bayes' theorem, with the posterior probabilities in the final column.

Alternatively, the prior probabilities and likelihoods can be entered in separate columns in your data matrix and multiplied by using TMUL. The posterior probabilities can then be found by using TDCN to divide the column created via TMUL by its sum.

SOLVED EXAMPLE

Screening tests are used in medicine to identify individuals who have, or are likely to have, certain diseases. The screening tests are by no means perfect, but the intent is usually to screen out those who are very unlikely to have a disease and to conduct further tests on those who have positive indications on the screening test. In the language of this chapter, the probability that a person has a disease is revised on the basis of the result of the screening test.

Suppose that a blood test is administered to all students at a large university. The blood test helps to identify students who might have a particular disease. Based on previous experience at this university and other universities, medical experts feel that the rate of incidence of the disease in the student population is about 3 percent. Thus, for any student, we have a prior probability of 0.03 that the student has the disease.

The blood test yields either a positive or negative reading. Its diagnostic accuracy can be summarized by two values: its false positive rate, which is 2 percent, and its false negative rate, which is 10 percent. The false positive rate of 2 percent means that

$$P(\text{positive}|\text{no disease}) = 0.02 \quad \text{and} \quad P(\text{negative}|\text{no disease}) = 0.98.$$

The false negative rate of 10 percent means that

$$P(\text{negative}|\text{disease}) = 0.10 \quad \text{and} \quad P(\text{positive}|\text{disease}) = 0.90.$$

What if a student receives a positive reading on the blood test? The posterior probability that the student has the disease is, from (B4–1),

$P(\text{disease}|\text{positive}) =$

$$\frac{P(\text{disease})P(\text{positive}|\text{disease})}{P(\text{disease})P(\text{positive}|\text{disease}) + [1\text{-}P(\text{disease})]P(\text{positive}|\text{no disease})}$$

$$= \frac{0.03(0.90)}{0.03(0.90) + 0.97\,(0.02)} = \frac{0.0270}{0.0270 + 0.0194}$$

$$= \frac{0.0270}{0.0464} = 0.58 \quad,$$

as shown in the output from SBAYE given in Figure B4.1. A positive reading increases the probability of having the disease from 0.03 to 0.58. A negative reading, on the other hand, reduces the probability from 0.03 to $P(\text{disease}|\text{negative}) = 0.003$, as shown in Figure B4.2.

The blood test might also provide information concerning other diseases. Suppose that we are interested in three diseases, which we shall call diseases D, E and F. A student either has D, E, or F or is healthy (denoted by H). (There is virtually no chance of having more than one of the three diseases.) The prior probabilities are

FIGURE B4.1 **Output from SBAYE for calculation of posterior probability following a positive reading in the medical screening example**

⟨n⟩ SBAYE

***** Revises Probabilities with Bayes' Theorem *****

Do you want to :
1. Revise probabilities for a single event
2. Revise probabilities for a number of events or classes
3. Revise a normal distribution for a mean
4. Revise a beta distribution for a proportion. (answer 1, 2, 3 or 4) ? 1

What is the prior probability for the event ? 0.03

What is the likelihood of the new information if the event occurs ? .90

What is the likelihood of the new information if the event does not occur ? 0.02

The posterior probability for the event is   0.582

FIGURE B4.2   Output from SBAYE for calculation of posterior probability following a negative reading in the medical screening example

⟨n⟩ SBAYE
$\ast\ast\ast\ast\ast$ Revises Probabilities with Bayes' Theorem $\ast\ast\ast\ast\ast$

Do you want to :
1. Revise probabilities for a single event
2. Revise probabilities for a number of events or classes
3. Revise a normal distribution for a mean
4. Revise a beta distribution for a proportion. (answer 1, 2, 3 or 4) ? 1

What is the prior probability for the event ? 0.03

What is the likelihood of the new information if the event occurs ? 0.10

What is the likelihood of the new information if the event does not occur ? 0.98

The posterior probability for the event is 0.003

$$P(D) = 0.03,$$
$$P(E) = 0.01,$$
$$P(F) = 0.04,$$
$$\text{and } P(H) = 0.92.$$

With a positive reading, the likelihoods are

$$P(\text{positive}|D) = 0.90,$$
$$P(\text{positive}|E) = 0.80,$$
$$P(\text{positive}|F) = 0.60,$$
$$\text{and } P(\text{positive}|H) = 0.02.$$

The calculation of posterior probabilities via SBAYE for this example is illustrated in Figure B4.3. Following a positive reading the probabilities are about 0.35, 0.10, and 0.31 of having diseases D, E, and F, and 0.24 of being healthy. The probabilities of having D, E, and F have increased considerably, and the probability of being healthy has gone down from 0.92 to 0.24.

EXERCISES

B4.1.
   In the blood test example involving only a single disease, suppose that the false positive rate is 8 percent and the false negative rate is 1 percent. Use SBAYE to find the probability that a student with a positive reading has the disease.

FIGURE B4.3 Output from SBAYE for calculation of posterior probabilities following a positive reading in the medical screening example with four categories (3 diseases and "healthy")

⟨n⟩ SBAYE
***** Revises Probabilities with Bayes' Theorem *****

Do you want to :
1. Revise probabilities for a single event
2. Revise probabilities for a number of events or classes
3. Revise a normal distribution for a mean
4. Revise a beta distribution for a proportion. (answer 1, 2, 3 or 4) ? 2

How many events or classes do you have? 4

What is the probability for the 1st event? .03

What is the probability for the 2nd event? .01

What is the probability for the 3rd event? .04

What is the probability for the 4th event? .92

What is the likelihood of the new information given the 1st event : .90

What is the likelihood of the new information given the 2nd event : .80

What is the likelihood of the new information given the 3rd event : .60

What is the likelihood of the new information given the 4th event : .02

Event	Prior Prob.	Likelihood	Pr. Prob*Lik.	Posterior Prob.
1	0.030	0.900	0.0270	0.349
2	0.010	0.800	0.0080	0.103
3	0.040	0.600	0.0240	0.310
4	0.920	0.020	0.0184	0.238
	1.000		0.0774	1.000

B4.2.

A banker feels that the probability that interest rates will decrease over the next year is 0.30, the probability they will increase is 0.60, and the probability they will remain the same is 0.10. A prominent economist then claims that interest rates will fall. The banker judges that the probability of this claim is 0.9 if interest rates will really decrease, 0.5 if they will actually increase, and 0.7 if they will remain the same. Use SBAYE to find the banker's posterior probabilities that rates will increase, decrease, or remain the same.

B4.3.

Compare the posterior probability obtained in Exercise B4.1 with the posterior probability in Figure B4.1. Can you explain the difference between the two numbers?

B5
Probability Distributions

When we are uncertain about the value of a variable (for example, the temperature tomorrow or the number that will come up in a toss of a die), we call it a random variable. Just as a set of data can be summarized by a frequency distribution, the uncertainty about a random variable can be summarized by a probability distribution, which consists of all possible values of the variable together with the associated probabilities (or cumulative probabilities in the case of a cumulative probability distribution). We focus here on discrete random variables, which have a countable number of possible values. Continuous random variables, which have an uncountable number of possible values, will be discussed in terms of certain special distributions such as the normal distribution (Chapter B9). Also, we will sometimes treat a continuous variable as if it is discrete, by rounding and considering only a countable number of values.

Probability distributions must obey two rules. First, no probability can be negative. Second, the sum of all the probabilities in a probability distribution must be one. Thus, we have

$$P(x) \geq 0 \text{ for all values } x \qquad \text{(B5–1)}$$

and

$$\sum_{\substack{\text{All values} \\ \text{of } x}} P(x) = 1 \quad . \qquad \text{(B5–2)}$$

An alternative to a listing of probabilities is a graph of a probability distribution. This is similar to a histogram, with the values of the random variable given on the horizontal axis and the probabilities represented by vertical bars.

ISP COMMANDS

Probability distributions can be entered in ISP by using ENTER twice, first for the values of the random variable and then for the probabilities. Commands involving special distributions such as the binomial and normal will be discussed in later chapters.

SOLVED EXAMPLE

Suppose that you are given an opportunity to play a game of chance. A fair die will be tossed and you will win or lose an amount depending on the number that comes up on the die:

> If 1 comes up, you lose $48.
> If 2 comes up, you lose $30.
> If 3 comes up, you lose $12.
> If 4 comes up, you don't win or lose anything.
> If 5 comes up, you win $48.
> If 6 comes up, you win $72.

Your gain from this game, which may be positive, negative, or zero, is a random variable. To display this probability distribution, ENTER was used to enter the six values -48, -30, -12, 0, 48, and 72. Then ENTER was used to enter the six probabilities, each of which is 1/6, or about 0.1667, and the display in Table B5.1 was obtained by using PRINT. The probabilities don't quite add up to one because of rounding error.

Various probabilities can be calculated from Table B5.1. For instance, the probability of a positive gain is

$$P(\text{GAIN}=48) + P(\text{GAIN}=72) = 0.1667 + 0.1667$$
$$= 0.333$$

TABLE B5.1   The probability distribution of the gain from a game of chance

Gain	Probability
−48.000	0.1667
−30.000	0.1667
−12.000	0.1667
0.000	0.1667
48.000	0.1667
72.000	0.1667
	1.0002

since 48 and 72 are the only positive gains that are possible. The cumulative probability $P(\text{GAIN} \leq -12)$ is

$$P(\text{GAIN} = -48) + P(\text{GAIN} = -30) + P(\text{GAIN} = -12)$$
$$= 0.1667 + 0.1667 + 0.1667 = 0.50.$$

EXERCISES

B5.1.

If a fair coin is tossed twice, find the probability distribution of the number of times heads comes up in the two tosses. If you will be paid $10 each time heads comes up and you will lose $6 each time tails comes up, find the probability distribution of your net gain from this coin-tossing game.

B5.2.

In a survey of American tourists traveling to Europe, 30 percent visit exactly 2 countries, 25 percent visit 3 countries, 20 percent visit 4 countries, 5 percent visit 5 countries, and none visit more than 5 countries. Find the probability distribution of the number of European countries visited by American tourists. What is the probability that a tourist will visit fewer than 4 countries?

B5.3.

The probability distribution of daily rainfall in inches in a particular city in July is $P(x=0) = 0.70$, $P(x=0.5) = 0.14$, $P(x=1) = 0.08$, $P(x=1.5) = 0.04$, $P(x=2) = 0.02$, $P(x=2.5) = 0.01$, and $P(x=3) = 0.01$. What is the probability of more than 1 inch of rain? What is the probability of between 1 and 2 inches of rain, inclusive?

B6

Expected Values

Data are often summarized by measures such as means and variances, as discussed in Chapters A3 and A4. In a similar fashion, probability distributions can be summarizied by various expected values, including the mean and variance. The mean of a probability distribution of a discrete variable, represented by Greek mu, can be found from the formula

$$\mu = \sum_{\substack{\text{All values} \\ \text{of } x}} x\, P(x) \quad . \tag{B6-1}$$

The mean tells us about an average value, while the variance tells us something about how spread out a probability distribution is. The variance of a probability distribution of a discrete variable, represented by Greek sigma squared, is

$$\sigma^2 = \sum_{\substack{\text{All values} \\ \text{of } x}} (x-\mu)^2 P(x) \quad . \tag{B6-2}$$

An alternative, computationally more efficient way to find the variance is

$$\sigma^2 = \sum_{\substack{\text{All values} \\ \text{of } x}} x^2 P(x) - \mu^2 \quad . \tag{B6-3}$$

As discussed in Chapter A4, a variance is expressed in squared units. To avoid squared units, we often work with the standard deviation, which is the square root of the variance and is denoted by Greek sigma. The interpretation of the mean, variance, and standard deviation of a random variable are the same as the interpretation of the same measures for a set of data, as presented in Chapters A3, A4, and A5. In fact, the formulas presented in this chapter are directly analogous to those presented in Chapter A5 for dealing with grouped data.

ISP COMMANDS

The formulas given in this chapter can be applied in ISP by using appropriate transformations. For example, you can enter the values of the random variable in one column of your data matrix and enter the probabilities in another column. The command TMUL can then be used to multiply these two columns, and the sum of the new column is the mean. For the variance, use TSCN to subtract the mean from each value of x, TCC to square these deviations, and TMUL to multiply the squared deviations by the probabilities. The sum of this last column is the variance.

However, there is an even easier way to find expected values via ISP. The command SEXPV will do all of the calculations and will print out the mean and the variance.

SOLVED EXAMPLE

An investor is faced with a choice between two investments. The first investment is thought of as relatively safe and conservative, while the second investment is viewed as more speculative and risky. The probability distributions of the returns from the two investments are shown in Table B6.1. You can see that Investment 2 has more extreme values, both on the good side (returns of 25 or 50 percent) and on the bad side (returns of zero or negative 25 percent). The best possible return for Investment 1 is only 20 percent, but the worst (5 percent) still represents a positive return. The mean and variance will help us summarize these probability distributions.

First, consider Investment 1. Entering the four possible returns (5, 10, 15, and 20) in the first column of the data matrix, entering the corresponding probabilities (.05, .30, .50, .15) in the second column, and using SEXPV yields the results shown in Figure B6.1. Investment 1 has an expected return, or mean return, of 13.75 percent with a variance of 14.6875 (which implies a standard deviation of 3.83).

Without using SEXPV, we could still find the mean and variance from formulas (B6–1) and (B6–2). This involves a number of transformations, which are summarized in Figure B6.2. The possible returns are labeled RET1, the probabilities PROB1, and the product of the returns and probabilities XPX (for $xP(x)$).The sum of the XPX column, 13.75, is the mean. Then 13.75 is subtracted (via TSCN) from each return in Column 1 to get DEV, the deviations from the mean. TCC squares the deviations to get DEVSQ,

TABLE B6.1 Probability distributions of percentage returns from two investments

Investment 1		Investment 2	
Return	Probability	Return	Probability
5	.05	−25	.20
10	.30	0	.20
15	.50	25	.30
20	.15	50	.30
	1.00		1.00

FIGURE B6.1   Use of SEXPV to find the mean and variance of return from Investment 1

⟨n⟩ SEXPV
                ***** Computes expected value *****
        Need help ? N
        Enter column number for random variable X ? 1
        Enter column number for probability ? 2
        Mean = 13.750000   , Variance = 14.687500

and TMUL is used to multiply the squared deviations and probabilities (Columns 5 and 2), yielding DSQPR. The sum of this last column is the variance, 14.6875.

What if you want to use formula (B6–3) to find the variance? The first three columns in Figure B6.3 are identical to those in Figure B6.2, and the fourth column lists the squared returns (using TCC on Column 1). Multiplying those squared returns by the probabilities (TMUL with Columns 4 and 2) gives the last column, the sum of which is

$$\sum_{\substack{\text{All values} \\ \text{of } x}} x^2 P(x) = 203.75 \quad .$$

From formula (B6–3), the variance is

$$\sigma^2 = 203.75 - (13.75)^2 = 14.6875 \quad .$$

FIGURE B6.2   Summary of ISP calculations of the mean and variance of return from Investment 1, using formulas (B6–1) and (B6–2)

⟨n⟩ PRINT

Observ.	RET1	PROB1	XPX	DEV	DEVSQ	DSQPR
1*	5.000	0.050	0.250	−8.750	76.563	3.828
2*	10.000	0.300	3.000	−3.750	14.063	4.219
3*	15.000	0.500	7.500	1.250	1.563	0.781
4*	20.000	0.150	3.000	6.250	39.063	5.859
Sum	50.000	1.0000	13.7500	−5.0000	131.2500	14.6875

Now we turn to Investment 2. As you can see from Figure B6.4, the results from applying SEXPV show that the mean return is 17.50 percent and that the variance is 756.25. The standard deviation is 27.50, the square root of the variance. Thus, we have the following comparison.

Investment 1: $\mu = 13.75$, $\sigma^2 = 14.6875$, $\sigma = 3.83$,
Investment 2: $\mu = 17.50$, $\sigma^2 = 756.25$, $\sigma = 27.50$.

The mean return of Investment 2 is 3.75 percent higher than that of Investment 1, but Investment 2 has a standard deviation more than seven times as large as that of

FIGURE B6.3   Summary of ISP calculations of the mean and variance of return from Investment 1, using formulas (B6–1) and (B6–3)

Observ.	RET1	PROB1	XPX	XSQ	XSQPR
1*	5.000	0.050	0.250	25.000	1.250
2*	10.000	0.300	3.000	100.000	30.000
3*	15.000	0.500	7.500	225.000	112.500
4*	20.000	0.150	3.000	400.000	60.000
Sum	50.0000	1.0000	13.7500	750.0000	203.7500

$$\sigma^2 = \Sigma x^2 P(x) - \mu^2 = 203.75 - (13.75)^2 = 14.6875$$

FIGURE B6.4   Use of SEXPV to find the mean and variance of return from Investment 2

⟨n⟩ SEXPV

***** Computes expected values *****

Need help? N
Enter column number for random variable X ? 1
Enter column number for probabilities ? 2

Mean = 17.50000   , Variance = 756.24994

Investment 1. In terms of the expected value, or the mean, Investment 2 is the more attractive of the two choices. However, an investor not willing to take much risk may prefer Investment 1 even though it has a lower mean. You can see that both the mean and the variance provide useful information in this example.

EXERCISES

B6.1.

For the probability distribution given in Exercise B5.3, use SEXPV to find the mean and variance of the amount of daily rainfall.

B6.2.

For the probability distribution given in Exercise B5.3, use ISP to find the mean and variance of the amount of daily rainfall *without* using SEXPV.

B6.3.

Find the mean and variance of the number of European countries visited by American tourists, using SEXPV with the information given in Exercise B5.2.

B6.4.

For the coin-tossing game in Exercise B5.1, what are the mean and variance of your net gain?

B7
The Binomial Distribution

The binomial distribution is applicable under the following conditions:

1. On each trial (that is, for each observation), there are two possible outcomes, which we will often refer to as success and failure.
2. The trials are independent.
3. The probability of success, denoted by p, remains the same for all trials.

If these conditions are met, then the distribution of x, the number of successes in n trials, is a binomial distribution. The formula for the binomial distribution is

$$P(x|n,p) = \frac{n!}{x! \, (n-x) \, !} \, p^x(1-p)^{n-x} \qquad \text{(B7–1)}$$

where x can be 0, 1, 2, and so on, up to n. Here $n!$ denotes "n factorial," which equals $n(n-1)(n-2) \cdots 2 \cdot 1$, and $x!$ and $(n-x)!$ are "x factorial" and "$n-x$ factorial," defined similarly. The mean of the binomial distribution can be expressed in the form

$$\mu = np \quad , \qquad \text{(B7–2)}$$

and the variance is

$$\sigma^2 = np(1-p) \quad . \qquad \text{(B7–3)}$$

The binomial distribution is quite useful, since many experiments consist of independent trials with two possible outcomes at each trial. Examples are yes/no answers, items that can be defective or good, patients who are cured or not, games that are won and lost, and so on.

ISP COMMANDS

Binomial probabilities can be found by using the ISP command DBINO. You can specify values for n, p, and x and have the computer provide the corresponding binomial probability. Or, you can have the computer print out a table of binomial probabilities; this is more useful than tables found in books, because you can specify values of p to more than two decimal places, the finest breakdown available in most printed tables. Finally, DBINO also will find the value of x corresponding to any cumulative probability for a binomial distribution (that is, for your choice of n and p).

The command PROBD can be used to display graphs of binomial distributions. Moreover, commands such as GCOIN, GDIE, and GMARB (with replacement) can provide illustrations of situations for which the binomial distribution is applicable. Data following a binomial model can also be generated via GSAMP, which will be discussed in detail in Part C.

If you want to try to use (B7–1) to calculate a binomial probability, the computing commands CFAC and CPOW will come in handy. CFAC calculates factorials, and CPOW computes expressions such as p^x. The number of sequences, or combinations, for a given n and x can be found by using GCOMB.

SOLVED EXAMPLE

A company manufactures large motors, and a rush order for 10 motors has just arrived. The inventory is totally depleted, so a production run is planned. In the past, 22 perecent of the motors produced have had defects requiring correction before the motors could be delivered. In this case there is no time for such corrections. Therefore, the production manager decides to produce 15 motors in order to be relatively confident of having 10 good ones to ship out right away.

Past history suggest that defective motors tend to occur randomly and independently. Thus, the binomial assumptions are satisfied. The distribution of x, the number of motors with defects in the lot of 15 motors, is a binomial distribution with $n = 15$ and $p = 0.22$. From formulas (B7–2) and (B7–3), the mean and variance of x are

$$\mu = 15(0.22) = 3.3 \quad \text{and} \quad \sigma^2 = 15(0.22)(0.78) = 2.574 \quad .$$

Thus, the expected number of defective motors in a lot of 15 is 3.3, with a standard deviation of

$$\sigma = \sqrt{2.574} = 1.604 \quad .$$

The company will be able to fill the rush order right away if there are at least 10 motors without defects in the lot of 15. This means that there must be at most 5 motors with defects. The probability of 5 or fewer defective motors in the lot of 15 is

$$P(x \leq 5) = P(x=0) + P(x=1) + P(x=2) + P(x=3) + P(x=4) + P(x=5) \quad .$$

The individual probabilities can be calculated from formula (B7–1). For example,

$$P(x=5) = \frac{15!}{5! \ 10!} (.22)^5 (.78)^{10} \quad .$$

Rather than perform the necessary calculations by hand, we can use DBINO, as illustrated in Figure B7.1. From these results, you can see that

$$P(x=5) = 0.1290 \quad .$$

But the probability that is needed is $P(x \leq 5)$. We could repeat DBINO for $x = 0$, 1, 2, 3, and 4. However, notice that the output in Figure B7.1 also includes

$$P(x < 5) = 0.7805 \quad .$$

Therefore,

$$P(x \leq 5) = P(x < 5) + P(x=5)$$
$$= 0.7805 + 0.1290 = 0.9095 \quad .$$

The probability of being able to fill the order without delay is 0.9095 if a production run of 15 motors is made.

Of course, instead of obtaining a single probability, we could look at a table of probabilities and see the entire distribution of x. DBINO can be used to get a table of

FIGURE B7.1   Use of DBINO to find binomial probabilities

```
⟨n⟩ DBINO
            ***** Binomial probabilities *****
Need help? N
Do you want :
1.   A table of binomial probabilities
2.   A single binomial probability
3.   The value of X corresponding to some cumulative prob.
Enter 1, 2, or 3 ? 2

Enter the number of trials : 15
Enter the probability of success: .22
Enter the value of the random variable X : 5
P(X= 5,n=15,p=0.220)  =  0.12900750
P(X< 5,n=15,p=0.220)  =  0.78053010
P(X> 5,n=15,p=0.220)  =  0.90462387E-01
```

binomial probabilities. If $n = 15$ and $p = 0.22$ are specified, a table will be given showing the binomial probabilities for $p = 0.02, 0.12, 0.32, 0.42,$ and 0.52 as well as $p = 0.22$. This output is presented in Figure B7.2. From the column headed 0.220 we see, for example, that the most likely number of defective motors is $x = 3$, with a probability of 0.2457. Values of x greater than 7 are quite unlikely.

FIGURE B7.2   Use of DBINO to display a table of binomial probabilities

Enter 1, 2, or 3 ? 1

Enter number of trials : 15

Enter the probability of success : .22

n = 15

x	.020	.120	p .220	.320	.420	.520
0	.7386	.1470	.0241	.0031	.0003	.0000
1	.2261	.3006	.1018	.0217	.0031	.0003
2	.0323	.2870	.2010	.0715	.0156	.0020
3	.0029	.1696	.2457	.1457	.0489	.0096
4	.0002	.0694	.2079	.2057	.1061	.0311
5	.0000	.0208	.1290	.2130	.1691	.0741
6	.0000	.0047	.0606	.1671	.2041	.1338
7	.0000	.0008	.0220	.1011	.1900	.1864
8	.0000	.0001	.0062	.0476	.1376	.2020
9	.0000	.0000	.0014	.0174	.0775	.1702
10	.0000	.0000	.0002	.0049	.0337	.1106
11	.0000	.0000	.0000	.0011	.0111	.0545
12	.0000	.0000	.0000	.0002	.0027	.0197
13	.0000	.0000	.0000	.0000	.0004	.0049
14	.0000	.0000	.0000	.0000	.0000	.0008
15	.0000	.0000	.0000	.0000	.0000	.0001

EXERCISES

B7.1.

Suppose that 20 percent of college graduates go on to graduate school. Use DBINO to find the probability that, among 15 randomly selected college students, exactly 3 will go to graduate school. What is the probability that more than 6 of the 15 will go to graduate school? What is the probability that 5 or fewer will go to graduate school?

B7.2.

Use DBINO to find the binomial distribution with $n = 10$ and $p = 0.37$. What are the mean and variance of this distribution?

B7.3.

Use GCOIN to simulate 100 tosses of 5 fair coins. What is the relative frequency of occurrence of 3 heads out of the 5 coins? From DBINO, what is the probability of obtaining 3 heads when 5 fair coins are tossed?

B7.4.

Use PROBD to display a graph of a binomial distribution with $n = 5$ and $p = 0.3$. Repeat the procedure with the same n but with $p = 0.5$ and $p = 0.7$.

B8
The Poisson Distribution

As in the case of the binomial, we can list conditions under which the Poisson distribution is applicable:

1. Occurrences of an event are observed over time, space, or some other continuum.
2. The occurrences are independent.
3. The event occurs randomly but at the same average rate, and the average rate is denoted by λ (Greek lambda).

For instance, the number of telephone calls received by the switchboard of a government agency between 10 a.m. and 11 a.m. and the number of particles of a pollutant in a cubic centimeter of air tend to behave like Poisson distributions.

If the Poisson conditions are met, then the distribution of x, the number of occurrences, is a Poisson distribution. The formula for the Poisson distribution is

$$P(x|\lambda) = \frac{\lambda^x e^{-\lambda}}{x!} \qquad \text{(B8–1)}$$

for $x = 0, 1, 2, 3, \dots$. The letter e stands for a mathematical constant, which represents the base of the natural logarithm system and equals 2.7183 (to 4 decimal places). You can think of e as being like π, another constant that you probably recall from geometry. As in the binomial formula, $x!$ stands for "x factorial," which equals $x(x-1)(x-2) \cdots 2 \cdot 1$. The mean and variance for a Poisson distribution are

$$\mu = \lambda \qquad \text{(B8–2)}$$

and

$$\sigma^2 = \lambda \quad . \qquad \text{(B8–3)}$$

If you want to find binomial probabilities when n is large, formula (B7–1) is difficult to apply (even on a computer) and binomial tables are not available for large n. However, when n is large and p is small (a rough rule of thumb is $n \geqslant 20$ and $p \leqslant 0.05$), the Poisson distribution can be used to approximate the binomial distribution. To use this approximation, simply let

$$\lambda = np \qquad\qquad\qquad (B8-4)$$

and calculate the Poisson probabilities.

ISP COMMANDS

Poisson probabilities can be found by using the ISP command DPOIS. You can specify values for λ and x and have the computer provide the corresponding Poisson probability. Or, you can have the computer print out a table of Poisson probabilities. This can be more useful than Poisson tables found in books, because you can specify values of λ to more decimal places than printed tables give. Finally, DPOIS will find the value of x corresponding to any cumulative probability for a Poisson distribution.

The command PROBD can be used to display graphs of Poisson distributions. Also, you can generate data that approximate a Poisson distribution by using GSAMP, which will be discussed in Part C.

SOLVED EXAMPLE

On a typical winter night, a heating contractor offering a 24-hour repair service averages one service call every two hours in the period between 9 p.m. and 7 a.m., and the occurrence of such calls seems to be random and independent. The cost of having someone on duty to handle service calls during this period will be covered only if three or more service calls are received. On any given night, what is the probability that the cost will be covered?

It appears that the Poisson model is appropriate, with service calls occurring randomly and independently over time at an average rate of one every two hours. In the ten-hour period from 9 a.m. to 7 p.m., the average number of calls is 10/2, or 5. Thus, the distribution of x, the number of service calls on any given night, is a Poisson distribution with $\lambda = 5$. From formulas (B8–2) and (B8–3), this distribution has mean

$$\mu = 5$$

and variance

$$\sigma^2 = 5 \quad .$$

The expected number of service calls in a night is 5, with a standard deviation of

$$\sigma = \sqrt{5} = 2.236 \quad .$$

FIGURE B8.1   Some output from DPOIS to find Poisson probabilities

```
⟨n⟩ DPOIS
        ***** Poisson distribution *****
Need help ? N
Do you want :
    1.  A table of Poisson probabilities
    2.  A single Poisson probability
    3.  The value of X corresponding to some cumulative prob.

Enter 1, 2, or 3 ? 2
Enter the value of the mean: lambda (np for binomial approx.) ? 5
Enter the value of the random variable X : 3
P(X= 3,lambda= 5.000) = 0.14037390
P(X< 3,lambda= 5.000) = 0.12465201
P(X> 3,lambda= 5.000) = 0.73497409
```

The probability that the cost will be covered is the probability that x is 3 or more. The probability that x is exactly 3 can be found from (B8–1) with $\lambda = 5$:

$$P(x=3) = \frac{5^3 e^{-5}}{3!} = 0.1404 \quad .$$

Figure B8.1 presents some output from DPOIS. The output includes $P(x=3) = 0.1404$ and also gives $P(x<3) = 0.1247$ and $P(x>3) = 0.7350$. Therefore, the probability we want is

$$P(x\geqslant 3) = 0.1404 + 0.7350 = 0.8754 \quad .$$

DPOIS can also be used to generate an entire table of probabilities, as illustrated in Figure B8.2. The Poisson distribution for $\lambda = 5$ is given, along with Poisson distributions for other nearby values of λ. From either Figure B8.1 or Figure B8.2, it can be determined that the probability of covering the cost of having someone on duty is 0.8754. The heating contractor can anticipate covering the cost 87.54 percent of the nights and not covering the cost the remaining 12.46 percent of the nights.

EXERCISES

B8.1.

Suppose that telephone calls arrive at a switchboard according to a Poisson distribution with $\lambda = 8$ per minute. Use DPOIS to find the probability that exactly 10 calls arrive in a particular minute. What is the probability that fewer than 6 calls arrive in a minute? How about the probability that 8 or more calls arrive in a minute?

FIGURE B8.2   Use of DPOIS to display a table of Poisson probabilities

Do you want :
1.   A table of Poisson probabilities
2.   A single Poisson probability
3.   The value of X corresponding to some cumulative prob.
Enter 1, 2, or 3, ? 1
Enter the value of the mean: lambda (np for binomial approx.) ? 5

x	4.600	4.700	4.800	4.900	5.000	5.100	5.200	5.300	5.400
0	.0101	.0091	.0082	.0074	.0067	.0061	.0055	.0050	.0045
1	.0462	.0427	.0395	.0365	.0337	.0311	.0287	.0265	.0244
2	.1063	.1005	.0948	.0894	.0842	.0793	.0746	.0701	.0659
3	.1631	.1574	.1517	.1460	.1404	.1348	.1293	.1239	.1185
4	.1875	.1849	.1820	.1789	.1755	.1719	.1681	.1641	.1600
5	.1725	.1738	.1747	.1753	.1755	.1753	.1748	.1740	.1728
6	.1323	.1362	.1398	.1432	.1462	.1490	.1515	.1537	.1555
7	.0869	.0914	.0959	.1002	.1044	.1086	.1125	.1163	.1200
8	.0500	.0537	.0575	.0614	.0653	.0692	.0731	.0771	.0810
9	.0255	.0281	.0307	.0334	.0363	.0392	.0423	.0454	.0486
10	.0118	.0132	.0147	.0164	.0181	.0200	.0220	.0241	.0262
11	.0049	.0056	.0064	.0073	.0082	.0093	.0104	.0116	.0129
12	.0019	.0022	.0026	.0030	.0034	.0039	.0045	.0051	.0058
13	.0007	.0008	.0009	.0011	.0013	.0015	.0018	.0021	.0024
14	.0002	.0003	.0003	.0004	.0005	.0006	.0007	.0008	.0009
15	.0001	.0001	.0001	.0001	.0002	.0002	.0002	.0003	.0003
16	.0000	.0000	.0000	.0000	.0000	.0001	.0001	.0001	.0001
17	.0000	.0000	.0000	.0000	.0000	.0000	.0000	.0000	.0000

B8.2.

   Use DPOIS to find the Poisson distribution with $\lambda = 13$. What are the mean and variance of this distribution?

B8.3.

   Use PROBD to display a graph of Poisson distributions with $\lambda = 4, 7,$ and 10.

B8.4.

   Suppose that 2 percent of the items produced in a given factory turn out to be defective. Use the Poisson approximation to the binomial distribution to find the probability that, in a lot of 400 items, no more than 10 are defective.

B9

The Normal Distribution

The normal distribution, which is the most widely used distribution in statistics, is a continuous distribution that looks like a bell-shaped curve. The "bell" is centered at the mean and is symmetric. As the standard deviation increases, the "bell" becomes more spread out. The normal curve, illustrated in Figure B9.1, has the formula

$$f(x) = \frac{1}{\sigma\sqrt{2\pi}}e^{-(x-\mu)^2/2\sigma^2} \text{ for } -\infty < x < \infty \quad , \qquad (B9-1)$$

where μ and σ are the mean and standard deviation, respectively. As noted in Chapter B8, e is a mathematical constant which is the base of the natural logarithm system and

FIGURE B9.1 A normal curve

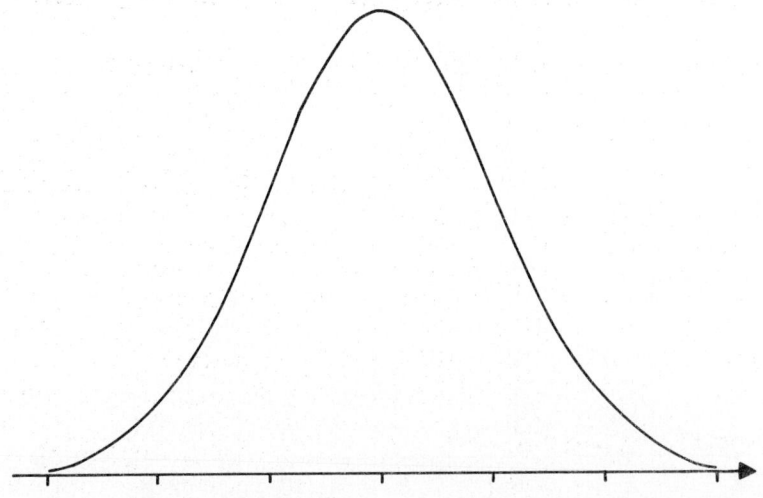

which is approximately 2.7183. In addition, formula (B9–1) contains another mathematical constant, π, which is approximately 3.14159. You probably recall π from geometry, where it is used in formulas for the area and circumference of a circle.

For continuous variables, probability is represented not by the height of the curve, but by *area* under the curve, as shown in Figure B9.2. In order to find areas under the normal curve, we first need to standardize, as in Chapter A7:

$$z = \frac{x - \mu}{\sigma} .$$

(B9–2)

The purpose of converting from x to z is that we then need a table for just one normal distribution, the standard normal distribution with mean zero and standard deviation one. Probabilities for z can be used to find probabilities for *any* normal distribution. Some types of probabilities available in tables of the normal distribution are shown in Figure B9.3. In calculating normal probabilities, we sometimes take advantage of the symmetry of the normal distribution, as illustrated in Figure B9.4.

As indicated in Chapter B8, the Poisson distribution can be used to approximate binomial probabilities when n is large and p is small. When n is large and p is not too small or too large [a rough rule of thumb is $np > 5$ *and* $n(1-p) > 5$], the normal distribution can be used to approximate the binomial by setting the mean equal to np and the variance equal to $np(1-p)$. To allow for the fact that the normal distribution is continuous and the binomial is discrete, we use a ''continuity correction'' before converting to z-values.

FIGURE B9.2   Probability as area under a normal curve

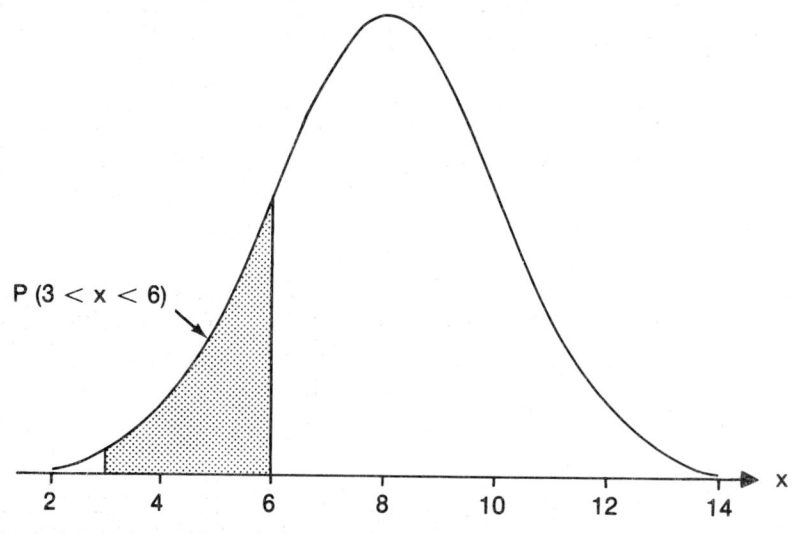

For example, $x = 39$ is converted to an interval, $38.5 \leqslant x \leqslant 39.5$. The normal distribution can also be used to approximate many other distributions.

ISP COMMANDS

Normal probabilities can be found by using the ISP command DNORM. You can specify values for μ, σ, and x and have the computer compute the standardized value and provide

FIGURE B9.3   Some probabilities from a standard normal distribution

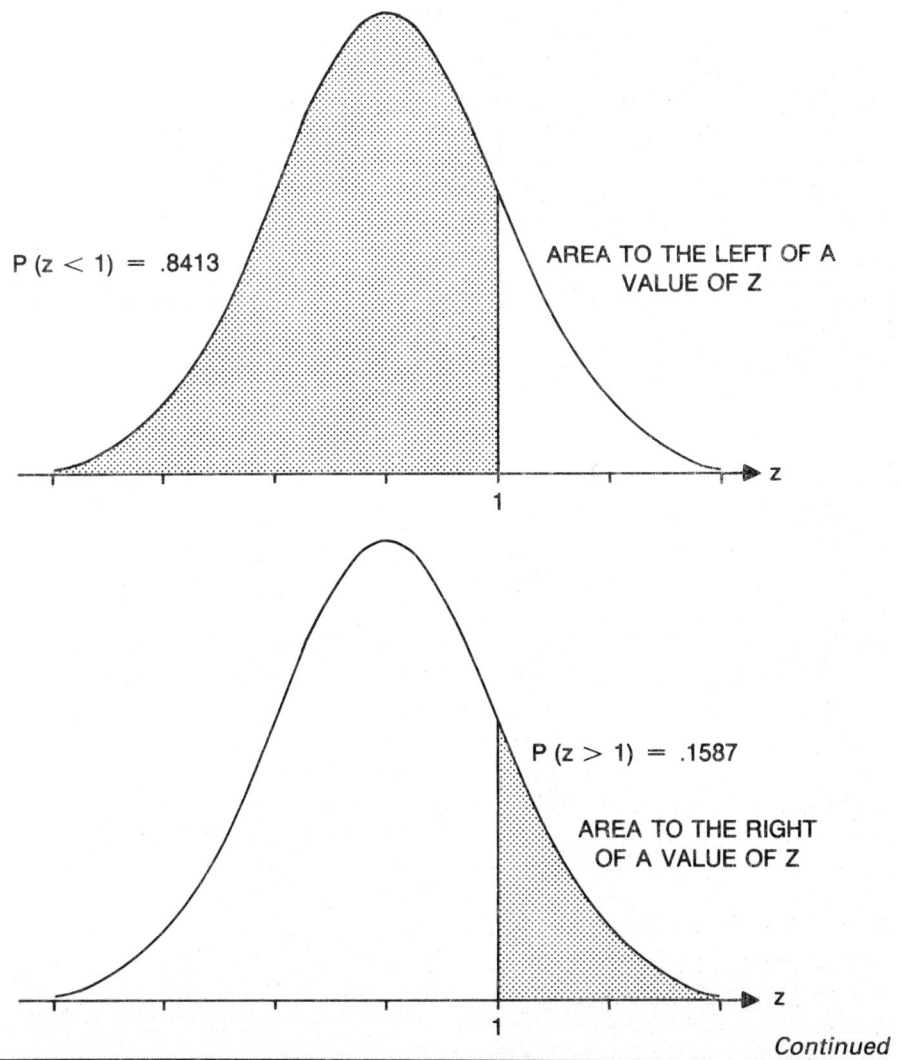

$P(z < 1) = .8413$

AREA TO THE LEFT OF A
VALUE OF Z

$P(z > 1) = .1587$

AREA TO THE RIGHT
OF A VALUE OF Z

Continued

FIGURE B9.3   Some probabilities from a standard normal distribution (cont.)

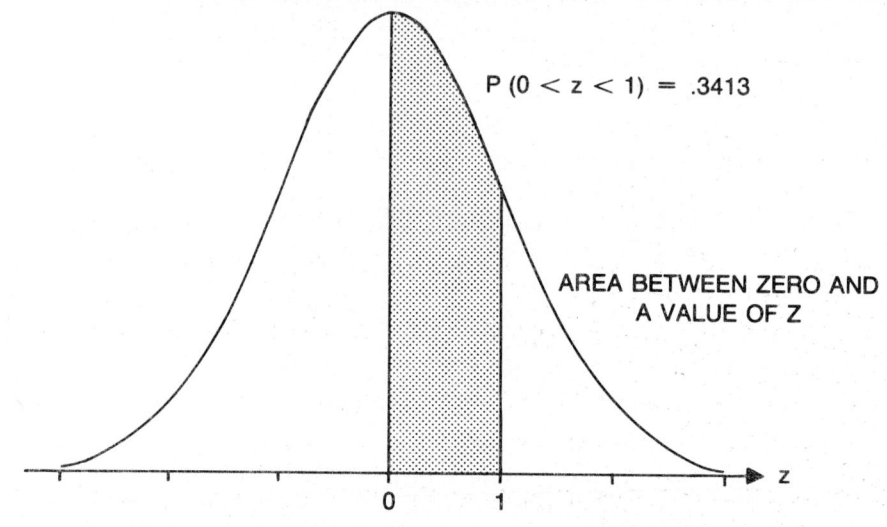

$P(0 < z < 1) = .3413$

AREA BETWEEN ZERO AND
A VALUE OF Z

FIGURE B9.4   The symmetry of a normal distribution

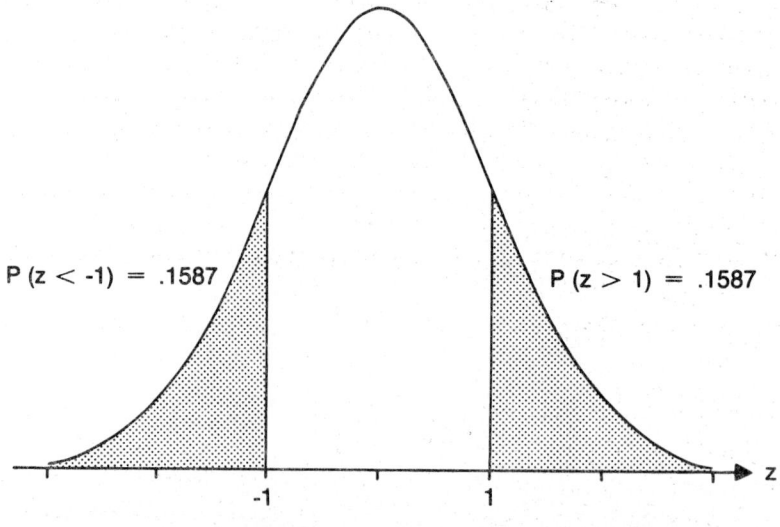

$P(z < -1) = .1587$

$P(z > 1) = .1587$

the three types of probability shown in Figure B9.3. DNORM also prints out a table of
standard normal probabilities. Moreover, DNORM can find the value of x corresponding
to a cumulative probability for any normal distribution (that is, for any choice of μ and σ).

The command PROBD can be used to display a graph of a normal distribution. Finally, if you would like to generate data that come from a normal distribution, use GSAMP.

SOLVED EXAMPLE

Suppose that an achievement test has been designed to produce scores that follow a normal curve with a mean score of 500 and a standard deviation of 100. This normal distribution is shown in Figure B9.5. If a student earns a score of 672 on the achievement test, how does this compare with other students? The score of 672 is above the mean, but how "good" is it?

To evaluate the score, let's find the probability of getting a better score. This is

$$P(x>672) = P\left(z>\frac{672-500}{100}\right)$$
$$= P(z>1.72) = 0.0427 \quad .$$

Here we standardized the score of 672 and then used Table B9.1, which was generated by DNORM. The result indicates that less than five percent of those who take the achievement test score higher than 672. An easier way to do this via DNORM is to let the computer do all the work, as illustrated in Figure B9.6.

How high a score would be necessary to be in the top one percent of those taking the test? DNORM can be used to answer this question, and the results are shown in

FIGURE B9.5   Distribution of scores on achievement test

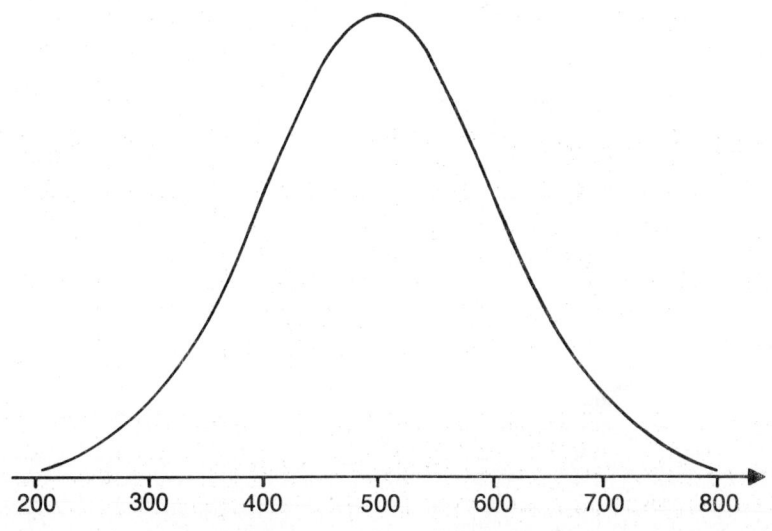

TABLE B9.1   A table of normal probabilities from DNORM

$\langle n \rangle$ DNORM

***** Normal Distribution *****

Need help? N

Do you want :
1. A table of normal probabilities
2. A single normal probability
3. The value of X corresponding to some cumulative prob.

Enter 1, 2, or 3 ? 1

Areas to the right of z value

z	.00	.01	.02	.03	.04	.05	.06	.07	.08	.09
0.0	.5000	.4960	.4920	.4880	.4840	.4801	.4761	.4721	.4681	.4641
.1	.4602	.4562	.4522	.4483	.4443	.4404	.4364	.4325	.4286	.4247
.2	.4207	.4168	.4129	.4090	.4052	.4013	.3974	.3936	.3897	.3859
.3	.3821	.3783	.3745	.3707	.3669	.3632	.3594	.3557	.3520	.3483
.4	.3446	.3409	.3372	.3336	.3300	.3264	.3228	.3192	.3156	.3121
.5	.3085	.3050	.3015	.2981	.2946	.2912	.2877	.2843	.2810	.2776
.6	.2743	.2709	.2676	.2643	.2611	.2578	.2546	.2514	.2483	.2451
.7	.2420	.2389	.2358	.2327	.2296	.2266	.2236	.2206	.2177	.2148
.8	.2119	.2090	.2061	.2033	.2005	.1977	.1949	.1922	.1894	.1817
.9	.1841	.1814	.1788	.1762	.1736	.1711	.1685	.1660	.1635	.1611
1.0	.1587	.1562	.1539	.1515	.1492	.1469	.1446	.1423	.1401	.1379
1.1	.1357	.1335	.1314	.1292	.1271	.1251	.1230	.1210	.1190	.1170
1.2	.1151	.1131	.1112	.1093	.1075	.1056	.1038	.1020	.1003	.0985
1.3	.0968	.0951	.0934	.0918	.0901	.0885	.0869	.0853	.0838	.0823
1.4	.0808	.0793	.0778	.0764	.0749	.0735	.0721	.0708	.0694	.0681
1.5	.0668	.0655	.0643	.0630	.0618	.0606	.0594	.0582	.0571	.0559
1.6	.0548	.0537	.0526	.0516	.0505	.0495	.0485	.0475	.0465	.0455
1.7	.0446	.0436	.0427	.0418	.0409	.0401	.0392	.0384	.0375	.0367
1.8	.0359	.0351	.0344	.0336	.0329	.0322	.0314	.0307	.0301	.0294
1.9	.0287	.0281	.0274	.0268	.0262	.0256	.0250	.0244	.0239	.0233
2.0	.0228	.0222	.0217	.0212	.0207	.0202	.0197	.0192	.0188	.0183
2.1	.0179	.0174	.0170	.0166	.0162	.0158	.0154	.0150	.0146	.0143
2.2	.0139	.0136	.0132	.0129	.0125	.0122	.0119	.0116	.0113	.0110
2.3	.0107	.0104	.0102	.0099	.0096	.0094	.0091	.0089	.0087	.0084
2.4	.0082	.0080	.0078	.0075	.0073	.0071	.0069	.0068	.0066	.0064
2.5	.0062	.0060	.0059	.0057	.0055	.0054	.0052	.0051	.0049	.0049
2.6	.0047	.0045	.0044	.0043	.0041	.0040	.0039	.0038	.0037	.0036
2.7	.0035	.0034	.0033	.0032	.0031	.0030	.0029	.0028	.0027	.0026
2.8	.0026	.0025	.0024	.0023	.0023	.0022	.0021	.0021	.0020	.0019
2.9	.0019	.0018	.0018	.0017	.0016	.0016	.0015	.0015	.0014	.0014
3.0	.0013	.0013	.0013	.0012	.0012	.0011	.0011	.0011	.0010	.0010

FIGURE B9.6   Use of DNORM to find normal probabilities

⟨n⟩ DNORM
          ***** Normal distribution *****

Need help? N
Do you want :
1.   A table of normal probabilities
2.   A single normal probability
3.   The value of X corresponding to some cumulative prob.

Enter 1, 2, or 3 ? 2
Enter the value of the mean : 500
Enter the value of std deviation or std error : 100
Enter the value of the random variable X : 672
The standardized Z value is : 1.7200
The cum. prob. from Z=0, to Z= 1.7200 is 0.4573
The cum. prob. from −infinity, to Z=1.7200 is 0.9573
The cum. prob. from Z= 1.7200 to +infinity is 0.0427

Figure B9.7. The dividing line for the top one percent is the point for which the area to the left under the curve is 0.99 and the area to the right under the curve is 0.01. Thus, in DNORM a right-hand cumulative probability of 0.01 is specified. The closest the normal table in DNORM comes to a right-tail area of 0.01 is 0.0099 at $z = 2.33$, which corresponds to

$$x = \mu + z\sigma = 500 + 2.33(100) = 733 \quad .$$

FIGURE B9.7   Use of DNORM to find a value of *x* corresponding to a given cumulative probability

⟨n⟩ DNORM
          ***** Normal distribution *****

Need help ? N
Do you want :
1.   A table of normal probabilities
2.   A single normal probability
3.   The value of X corresponding to some cumulative prob.

Enter 1, 2, or 3 ? 3
Enter the value of the mean : 500
Enter the value of std deviation or std error : 100
Enter the right-hand cumulative probability : .01
For a right-hand cum. prob. of 0.0099 a mean of 500.0000 and a std dev. (or error) or 100.0000 the corresponding X value is 733.

This means that a score of 733 would put a student in the top one percent.

EXERCISES

B9.1.

The time it takes a student to go from her house to school in the morning is normally distributed with a mean of 35 minutes and a standard deviation of 4 minutes. Use DNORM to find the probability that she arrives at school in less than 28 minutes. If she leaves her house 40 minutes before the start of her first class, what is the probability that she will be late for class? What is the probability of being late if she leaves 45 minutes before the first class?

B9.2.

In Exercise B9.1, when should the student leave home in order to have exactly a 0.90 probability of arriving at school before the class starts?

B9.3.

Use PROBD to display a graph of a normal distribution with a mean of 50 and a standard deviation of 15.

B9.4.

If the maximum temperature in June in New York is normally distributed with a mean of 76 degrees Fahrenheit and a standard deviation of 7 degrees, what is the probability of a maximum temperature of 90 or higher? How low is the maximum temperature on the coolest 10% of the days?

Summary of Part B

In everyday life as well as in complex, serious, large-scale problems, we are faced with uncertainty. In order to make better, more intelligent inferences and decisions, we have to be able to deal with uncertainty. Probability is the language of uncertainty. The purpose of Part B has been to introduce you to this language and some tools for working with it, thereby helping you to understand and measure uncertainty.

Probabilities for events can be interpreted in terms of equally likely outcomes, relative frequencies, or subjective judgments. Regardless of the interpretation, the probabilities must obey certain rules. Furthermore, in addition to probabilities for events, we can talk about probability distributions for random variables. Such distributions and their summary measures are analogous to the frequency distributions and summary measures discussed in Part A. Certain distributions are of particular interest in probability and statistics, and three of these are the binomial, Poisson, and normal distributions, with the normal distribution being the most widely used distribution in statistics. The knowledge of some basic concepts of probability and probability distributions, and some familiarity with important distributions such as the normal, will be very helpful in the study of inferential statistics.

VOCABULARY LIST

---

addition rule for mutually exclusive events

addition rule for non-mutually exclusive events

area under curve as probability

Bayes' theorem

binomial distribution

binomial probability

classical interpretation of probability

complementary event

conditional probability

continuity correction

continuous random variable

cumulative probability

cumulative probability distribution

dependent events

discrete random variable

equally likely events

event

expected value

experiment

independence

independent events

independent trials

independent variables

inferential statistics

joint probability

law of large numbers

likelihood

marginal probability

mean

multiplicative rule for dependent events

multiplicative rule for independent events

mutually exclusive events

normal distribution

normal probability

odds

outcome

Poisson distribution

Poisson probability

posterior probability

prior probability

probability

probability distribution

random variable

relative frequency interpretation of probability

revision of probabilities

rules of probability

sample space

standard deviation

standard normal distribution

subjective interpretation of probability

trial

uncertainty

variance

LIST OF SYMBOLS

P — probability

A, B, etc. — events

$P(A)$ — the probability that event A occurs

$P(A$ or $B)$ — the probability that either A or B occurs (or both occur)

$P(A$ and $B)$ — the probability that both A and B occur

$P(A|B)$ — the conditional probability of A, given B

x, y, etc. — random variables

$P(x)$ — the probability distribution for a discrete random variable x

$P(x>c)$ — the probability that the random variable x is greater than c

$P(x<c)$ — the probability that the random variable x is less than c

$P(a<x<b)$ — the probability that the random variable x is between a and b

μ — the mean of a probability distribution (or of a population when the probability distribution represents a population)

σ^2 — the variance of a probability distribution (or of a population)

σ — the standard deviation of a probability distribution (or of a population)

n	– sample size (the number of trials)
p	– the probability of success in a binomial distribution
λ	– the average rate of occurrence in a Poisson distribution
e	– the base of the natural logarithm system, a mathematical constant, approximately 2.7183
$f(x)$	– the curve representing the probability distribution for a continuous random variable x
π	– the ratio of the circumference to the diameter of a circle, a mathematical constant, approximately 3.14159
z	– a standardized variable
$n!, x!$, etc.	– n factorial $[n(n-1)(n-2)\cdots 1]$, x factorial $[x(x-1)(x-2)\cdots 1]$, etc.

FORMULAS

Odds in favor of A $= \dfrac{P(A)}{1 - P(A)}$

$P(A) = \dfrac{\text{Odds in favor of } A}{1 + \text{Odds in favor of } A}$

$P(A) \geqslant 0$
$P(A) \leqslant 1$
$P(A \text{ or } B) = P(A) + P(B)$ if A and B are mutually exclusive
$P(A \text{ or } B) = P(A) + P(B) - P(A \text{ and } B)$
$P(\text{not } A) = 1 - P(A)$
$P(A \text{ and } B) = P(A)P(B)$ if A and B are independent
$P(A \text{ and } B) = P(A)P(B|A)$

$P(B|A) = \dfrac{P(A \text{ and } B)}{P(A)}$

$P(B|A) = P(B)$ if A and B are independent
$P(A, B, \text{ and } C) = P(A)P(B|A)P(C|A \text{ and } B)$

$P(B|A) = \dfrac{P(B)P(A|B)}{P(B)P(A|B) + [1-P(B)]P(A|\text{not}B)}$

$P(B_i|A) = \dfrac{P(B_i)P(A|B_i)}{\displaystyle\sum_{j=1}^{k} P(B_j)P(A|B_j)}$ if $B_1, \ldots, B_k$ are mutually exclusive and exhaust all possibilities

$P(x) \geqslant 0$

$$\sum_{\substack{\text{All values} \\ \text{of } x}} P(x) = 1$$

$$\mu = \sum_{\substack{\text{All values} \\ \text{of } x}} x P(x) \qquad \text{(for a discrete random variable)}$$

$$\sigma^2 = \sum_{\substack{\text{All values} \\ \text{of } x}} (x - \mu)^2 P(x) \qquad \text{(for a discrete random variable)}$$

$$\sigma^2 = \sum_{\substack{\text{All values} \\ \text{of } x}} x^2 P(x) - \mu^2 \qquad \text{(for a discrete random variable)}$$

$$\sigma = \sqrt{\sum_{\substack{\text{All values} \\ \text{of } x}} (x - \mu)^2 P(x)} \qquad \text{(for a discrete random variable)}$$

$$\sigma = \sqrt{\sum_{\substack{\text{All values} \\ \text{of } x}} x^2 P(x) - \mu^2} \qquad \text{(for a discrete random variable)}$$

$$P(x|n,p) = \frac{n!}{x!(n-x)!} p^x (1-p)^{n-x} \qquad \text{(binomial distribution)}$$

$\mu = np \qquad \text{(for a binomial distribution)}$

$\sigma^2 = np(1-p) \qquad \text{(for a binomial distribution)}$

$$P(x|\lambda) = \frac{\lambda^x e^{-\lambda}}{x!} \qquad \text{(Poisson distribution)}$$

$\mu = \lambda \qquad \text{(for a Poisson distribution)}$

$\sigma^2 = \lambda \qquad \text{(for a Poisson distribution)}$

$\lambda = np \qquad$ (for the Poisson approximation to a binomial distribution)

$$f(x) = \frac{1}{\sigma\sqrt{2\pi}} e^{-(x-\mu)^2/2\sigma^2} \qquad \text{(normal distribution)}$$

$$z = \frac{x - \mu}{\sigma}$$

RELEVANT ISP COMMANDS

For Part B, ISP is useful not just in a computational sense, but also in terms of providing some experience with uncertainty and random fluctuations through the G commands. The following commands are helpful in this regard.

GCARD: Simulates a deck of playing cards
GCOIN: Simulates the tossing of coins
GDIE: Simulates the throwing of dice
GLOTO: Simulates the game of LOTTO
GMARB: Simulates sampling marbles from an urn
GROUL: Simulates the game of roulette
GTACK: Simulates the tossing of a thumbtack

Another G command that will be given more emphasis in Part C is GSAMP, which generates single or repeated samples from a data set or a theoretical distribution.

As far as computations are concerned, the C and T commands listed at the end of Part A can be helpful in Part B as well. In addition, certain commands are designed especialy with some of the computations required in Part B in mind. The command SBAYE can be used to revise probabilities via Bayes' theorem. Another S command, SEXPV, provides the mean and variance of a probability distribution.

The three distributions covered in the last three chapters of Part B are treated in individual commands:

DBINO: Binomial distribution
DPOIS: Poisson distribution
DNORM: Normal distribution

These three commands can be used to generate tables of probabilities, to provide probabilities for given values of the random variable, and to provide values of the random variable corresponding to given probabilities. The command PROBD can plot these three distributions as well as others, and PDIST can be used to compare theoretical distributions such as these with frequency distributions of data.

The topics covered in the nine chapters of Part B are summarized briefly in the nine MP modules.

MPB1:   Introduction to probability
MPB2:   Rules of probability
MPB3:   Conditional probability

PART C
Sampling and Sampling Distributions

Often we cannot obtain all the data we would ideally like to have about a population. When it is impractical or even impossible to get data concerning every member of the population, we may be able to take a sample that consists of a portion of the population. We can then use the information from the sample to make inferences about the entire population. Although the sample information may be less information than we ideally want, it can still be very valuable in helping us to make better, more informed decisions.

By its very nature, a sample does not tell us everything about the population. The sample results will depend on exactly which members of the population are chosen. With a random sampling plan, we can determine how much variability to expect among samples. Our uncertainty about a sample mean or any other sample statistic can be expressed as a probability distribution called a sampling distribution. The notion of a sampling distribution of a statistic tells us how much that statistic is likely to vary just by chance (the "luck of the draw"), and this in turn tells us how accurate inferences based on the statistic are likely to be.

Chapter C1 introduces the idea of sampling, with emphasis on simple random sampling. The general idea of a sampling distribution is introduced in Chapter C2. A specific sampling distribution, the sampling distribution of the mean, is discussed in Chapters C3 and C4. Finally, Chapter C5 indicates how the concept of sampling distributions can be used in statistical inference and thus provides a preview of the material on statistical inference in Parts D, E, and F.

C1

Sampling

Because of the time, trouble, and expense associated with examining an entire population, data are usually collected for only a portion of the population, called a sample. To avoid systematic biases in inference, statisticians generally choose samples by using random sampling. Simple random sampling is a sampling plan in which every possible sample of n members of the population has the same chance of being selected. A "fair" sampling plan such as this provides no guarantee that the sample will be representative of the population; an unrepresentative sample may result by the luck of the draw. However, systematic biases are avoided. Simple random sampling is a probability sampling plan because each possible sample has a known probability of being selected. Other probability sampling plans, such as stratified random sampling (dividing a population into groups and using random sampling separately for each group), can also be used.

Simple random samples can be selected by using some physical device such as lottery tickets, but it is easier to use tables of random numbers or simply let a computer perform the random selection. In selecting a sample, it is important to distinguish between sampling without replacement (allowing a member of the population to be included in the sample at most once) and sampling with replacement (permitting inclusion more than once). The distinction becomes less important when the sample is only a small portion of the population.

ISP COMMANDS

The command GSAMP enables you to select a simple random sample from any data set. The data can be entered at the terminal or taken from a file. The sampling can be done with or without replacement. GSAMP can also provide samples from populations represented by theoretical distributions. For example, you can take a sample from a normal population with a mean of 100 and a standard deviation of 20 (or any mean and standard deviation you wish to use), a uniformly distributed population, or populations with Poisson or gamma distributions. In GSAMP, you can have the computer print a frequency distribution and a histogram for both the population and the sample.

The "games" available on ISP also illustrate random sampling. For example, GLOTO involves the sampling of lottery numbers, GMARB involves the sampling of marbles, and cards are sampled from a deck of cards in GCARD. The commands GCOIN, GTACK, GDIE, and GROUL demonstrate sampling from processes such as coin tossing.

You may want to go through the details of taking a sample yourself instead of leaving all of the work to the computer. Random numbers will be needed to select a simple random sample. Random numbers in any specified interval can be generated in ISP by using the DUNIF command.

SOLVED EXAMPLE

The ages of 230 students in an MBA program are presented in Table C1.1. If we did not have these data from our population of 230 students, we might obtain information about the ages of students in the program by taking a sample. To see how this could be done, we can use GSAMP to take a sample from the population of ages.

First, suppose that we take a random sample of size 15 with replacement. The computer selects such a sample from a population of 230 by generating 15 random numbers between 1 and 230. A sample of size 15 generated by using GSAMP is shown in Table C1.2. The first random number generated by the computer is 91. Thus, the 91st element of the population is the first member of the sample. From Table C1.1, the 91st student is 26 years old; there are 20 ages in each row, so that the 91st age is the 11th age in the fifth row. The second random number is 88, corresponding to an age of 25 (the 8th age in the 5th row). The third random number is 124, corresponding to an age of 27 (the 4th age in the 7th row). Continuing in this fashion, we wind up with a random sample of size 15.

When sampling is done with replacement, it is possible for the same random number and therefore the same element of the population to appear more than once. In the sample shown in Table C1.2, the 91st student is included twice in the sample because the random

TABLE C1.1   A population of ages of 230 students

27	25	25	32	30	29	26	28	28	26	21	31	35	25	27	32	28	24	29	31
32	29	27	27	24	24	27	27	26	28	26	26	28	22	29	29	26	28	29	24
28	25	28	33	28	28	29	30	27	26	24	30	25	35	28	26	28	31	30	36
26	29	26	20	31	28	29	28	24	28	26	30	29	29	30	28	31	23	28	30
28	29	27	33	27	23	25	25	30	32	26	30	31	27	24	27	28	26	27	28
25	31	32	22	31	29	26	26	26	27	29	29	28	30	26	30	33	27	28	27
29	24	26	27	28	26	29	24	29	27	29	33	28	28	30	30	27	26	28	29
31	26	26	30	25	31	28	27	25	23	27	31	27	27	26	29	27	28	30	25
26	23	26	28	25	28	26	31	24	27	25	25	27	28	26	26	27	30	26	30
31	32	26	31	25	28	27	28	36	27	28	23	29	27	29	25	30	30	22	26
27	27	30	31	27	26	23	28	28	28	23	29	32	26	25	27	27	29	29	25
27	38	34	34	29	29	25	37	34	29										

number 91 happened to come up twice in the 15 random numbers. If we instruct the computer to sample without replacement when we use GSAMP, this repetition cannot occur. The results of a random sample of size 15 taken without replacement from the population of ages are given in Table C1.3. Note that no students are included more than once in the sample. Also, note that the first student in the sample is the 133rd student,

TABLE C1.2   A random sample of size 15 with replacement from the population of ages

Element sampled	Population position	Value
1	91	26.00
2	88	25.00
3	124	27.00
4	203	30.00
5	95	24.00
6	184	31.00
7	101	25.00
8	6	29.00
9	32	26.00
10	156	29.00
11	91	26.00
12	69	24.00
13	43	28.00
14	85	27.00
15	32	26.00

TABLE C1.3   A random sample of size 15 without replacement from the population of ages

Element sampled	Population position	Value
1	133	28.00
2	200	26.00
3	201	27.00
4	92	30.00
5	151	27.00
6	69	24.00
7	65	31.00
8	79	28.00
9	33	28.00
10	53	25.00
11	220	25.00
12	23	27.00
13	82	29.00
14	180	30.00
15	90	32.00

not the 91st student. In GSAMP, a new set of random numbers is generated for each sample.

EXERCISES

C1.1.

Data concerning the number of children in the families in a small French community are given in Table A1.6. Enter this data set in ISP, and use GSAMP to take a sample of 10 families (without replacement).

C1.2.

Use GSAMP to take simple random samples of size 10 with and without replacement from the data in Table A1.5 concerning the number of accidents reported at a busy intersection. Is the distinction between sampling with replacement and sampling without replacement important in this situation?

C1.3.

Scores on an aptitude test are normally distributed with mean 500 and standard deviation 100. Use GSAMP to generate a sample of 20 scores from this distribution of scores.

C1.4.

Cars arrive at a toll booth according to a Poisson distribution with $\lambda = 4$ per minute. Use GSAMP to sample from a Poisson distribution with $\lambda = 4$ in order to simulate the arrival of cars at the toll booth. Take 10 samples, representing 10 minutes. Does the number of cars arriving vary from sample to sample (minute to minute)? Explain why such differences might occur.

C2
Sampling Distributions

The various samples of a given size that might be obtained from simple random sampling differ from each other because the members of the population that are included in a particular sample are a matter of chance (the chance associated with random sampling). The differences among possible samples are called sampling fluctuations. In making inferences from samples, we need to know how large the sampling fluctuations are likely to be. This is what a sampling distribution tells us. The sampling distribution of a sample mean, for example, is the probability distribution showing how likely various values of the sample mean are. We also work with sampling distributions for statistical measures other than the sample mean. An understanding of the concept of a sampling distribution is central to an understanding of the basic notions of statistical inference.

ISP COMMANDS

The command GSAMP can be used to generate repeated random samples from a data set or a theoretical distribution. After the first sample is displayed, GSAMP asks how many additional samples (of the same size) are wanted. The sample mean and sample standard deviation are computed for each sample. Moreover, a frequency distribution and histogram of the sample means can be printed. This provides an estimate of the theoretical sampling distribution of the sample mean.

In commands such as GCOIN, GDIE, and GCARD, repeated sampling is possible. The resulting frequency distributions of sample statistics are compared with the actual theoretical sampling distributions.

SOLVED EXAMPLE

Consider once again the population of ages of 230 students given in Table C1.1. A histogram generated by using the command PHIST on this set of data is shown in Figure C2.1. Recall the random sample of $n = 15$ ages displayed in Table C1.2. We could

FIGURE C2.1   Histogram from PHIST of population of ages of 230 students

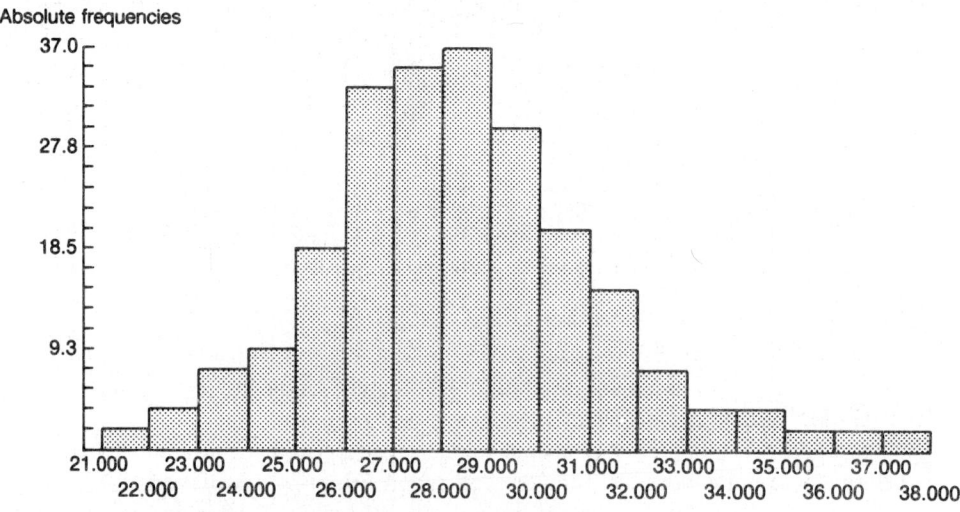

Population mean: 27.839130   Standard deviation: 2.8492403

compute the sample mean and sample standard deviation for this sample. If we took another sample of size 15 from the same population, we would expect to find a different sample mean and sample variance because of sampling fluctuations.

Table C2.1 shows the sample means and sample standard deviations of 200 separate random samples of $n = 15$ ages generated by GSAMP from our population of ages. A frequency distribution and histogram of the 200 sample means are displayed in Figure C2.2, and a frequency distribution and histogram of the 200 sample standard deviations are given in Figure C2.3. These displays give an indication of the variability in $\bar{x}$ and s from samples of size $n = 15$ from this population. They provide rough estimates of the sampling distributions of $\bar{x}$ and s.

Figures C2.2 and C2.3 represent only rough estimates of the sampling distributions of $\bar{x}$ and s because they are based on only 200 samples of size 15. A larger number of repeated samples would give better estimates. For instance, the frequency distribution and histogram of $\bar{x}$-values presented in Figure C2.4 are based on the sample mean ages for 2501 random samples of size 15 generated by using GSAMP.

TABLE C2.1   Sample means and standard deviations of 200 samples of _n_ = 15 ages generated by GSAMP

Sample	Sample Mean	Standard Deviation
1	26.53	2.67
2	27.53	1.55
3	28.13	1.88
4	27.07	2.58
5	28.50	3.02
6	29.07	3.03
7	28.27	1.98
8	26.80	2.21
9	28.40	3.07
10	27.80	1.97
11	27.87	2.56
12	28.73	2.99
13	27.87	3.34
14	28.20	1.42
15	28.73	3.08
16	28.67	3.35
17	27.73	3.77
18	27.47	3.25
19	28.40	2.35
20	27.80	2.62
Do you want more printing ? Y		
21	28.13	2.59
22	27.67	3.46
23	28.80	3.38
24	26.27	1.58
25	27.27	3.86
26	26.60	2.72
27	26.93	2.46
28	27.60	1.88
29	27.67	2.85
30	27.80	2.81
31	26.53	2.23
32	28.13	3.64
33	27.80	2.18
34	28.07	2.34
35	27.87	2.53
36	27.27	1.79
37	27.20	2.37
38	28.13	2.42
39	26.53	2.23
40	26.67	2.19
Do you want more printing ? Y		
41	27.00	3.72
42	27.13	2.97
43	28.20	3.55
44	27.93	4.70

Continued

TABLE C2.1   Sample means and standard deviations of 200 samples of n = 15 ages generated by GSAMP (cont.)

Sample	Sample Mean	Standard Deviation
45	28.80	3.82
46	28.33	3.09
47	28.33	3.18
48	27.40	2.32
49	27.20	2.60
50	27.67	2.09
51	27.13	2.90
52	29.13	3.74
53	28.67	2.55
54	28.07	3.41
55	28.33	3.66
56	27.40	1.96
57	27.40	2.41
58	29.00	4.21
59	28.27	2.55
60	28.47	2.85
Do you want more printing ? Y		
61	26.87	1.25
62	28.07	2.37
63	27.80	3.61
64	29.07	3.49
65	27.47	2.42
66	27.07	2.28
67	28.00	3.21
68	27.80	2.88
69	26.87	2.92
70	27.07	3.26
71	28.00	3.63
72	28.47	3.16
73	27.47	1.77
74	28.20	2.46
75	27.13	3.76
76	27.40	3.07
77	26.73	3.24
78	28.27	2.40
79	28.20	3.32
80	28.07	2.49
Do you want more printing ? Y		
81	28.33	2.19
82	26.60	2.59
83	27.13	2.20
84	28.07	3.22
85	27.87	2.03
86	27.73	2.37
87	28.00	2.88
88	27.47	2.80

Continued

TABLE C2.1 Sample means and standard deviations of 200 samples of n = 15 ages generated by GSAMP (cont.)

Sample	Sample Mean	Standard Deviation
89	27.80	1.82
90	29.07	2.81
91	29.00	3.42
92	28.40	4.27
93	27.93	1.33
94	28.87	3.66
95	28.93	3.10
96	28.20	3.73
97	28.20	3.12
98	28.87	4.17
99	28.20	2.54
100	26.20	2.04
Do you want more printing ? Y		
101	27.93	2.89
102	28.47	3.72
103	26.47	2.47
104	27.47	2.95
105	28.27	3.31
106	27.47	2.64
107	27.20	2.78
108	29.13	3.44
109	27.60	1.72
110	28.47	3.54
111	27.40	3.18
112	28.27	2.49
113	28.20	3.59
114	29.40	3.04
115	28.20	2.98
116	26.07	2.25
117	27.40	2.38
118	27.00	4.49
119	27.93	4.32
120	29.13	2.13
Do you want more printing ? Y		
121	27.47	2.17
122	28.13	3.00
123	27.13	1.73
124	28.13	2.92
125	27.20	1.32
126	26.93	1.91
127	28.00	3.57
128	28.13	3.27
129	28.27	2.31
130	26.80	2.46
131	27.80	2.57
132	27.53	2.47

Continued

TABLE C2.1 Sample means and standard deviations of 200 samples of n = 15 ages generated by GSAMP (cont.)

Sample	Sample Mean	Standard Deviation
133	29.73	3.97
134	28.73	1.94
135	27.00	2.36
136	27.87	2.36
137	27.47	3.52
138	28.60	2.72
139	26.93	4.08
140	28.13	3.44
Do you want more printing ? Y		
141	26.87	1.96
142	27.80	3.47
143	27.87	2.56
144	27.47	2.03
145	27.80	2.93
146	26.67	2.79
147	28.47	2.97
148	27.20	3.19
149	27.53	2.95
150	28.07	3.45
151	27.80	2.54
152	27.80	2.93
153	27.87	4.00
154	27.20	3.95
155	27.33	2.35
156	27.60	1.84
157	28.80	3.61
158	27.13	1.36
159	26.87	2.53
160	28.20	2.24
Do you want more printing ? Y		
161	27.27	2.22
162	29.13	4.26
163	28.07	2.19
164	27.53	2.53
165	27.27	2.43
166	28.47	1.73
167	28.53	3.85
168	27.93	3.39
169	28.60	4.61
170	27.33	1.59
171	28.00	3.07
172	29.73	3.53
173	26.33	2.26
174	27.67	4.05
175	27.27	2.43
176	28.07	3.22

Continued

TABLE C2.1 Sample means and standard deviations of 200 samples of $n = 15$ ages generated by GSAMP (cont.)

Sample	Sample Mean	Standard Deviation
177	28.00	2.85
178	28.47	2.45
179	28.07	2.49
180	28.07	3.65
Do you want more printing ? Y		
181	28.33	3.11
182	28.00	2.56
183	27.80	1.93
184	29.53	3.40
185	27.73	2.69
186	28.47	2.56
187	27.20	1.57
188	27.13	1.51
189	27.00	3.66
190	26.73	1.94
191	27.27	3.28
192	26.33	2.55
193	27.33	2.23
194	27.60	2.20
195	27.20	3.00
196	27.27	2.58
197	28.27	3.37
198	27.93	3.35
199	28.87	3.83
200	29.07	3.20

FIGURE C2.2 Frequency distribution and histogram for sample means from 200 samples of n = 15 ages

Values of $\bar{x}$	Absolute frequency	Relative frequency
26.00–26.29	3	0.015
26.30–26.59	6	0.030
26.60–26.89	12	0.060
26.90–27.19	17	0.085
27.20–27.49	33	0.165
27.50–27.79	15	0.075
27.80–28.09	43	0.215
28.10–28.39	30	0.150
28.40–28.69	17	0.085
28.70–28.99	10	0.050
29.00–29.29	10	0.050
29.30–29.59	2	0.010
29.60–29.89	2	0.010
	200	1.000

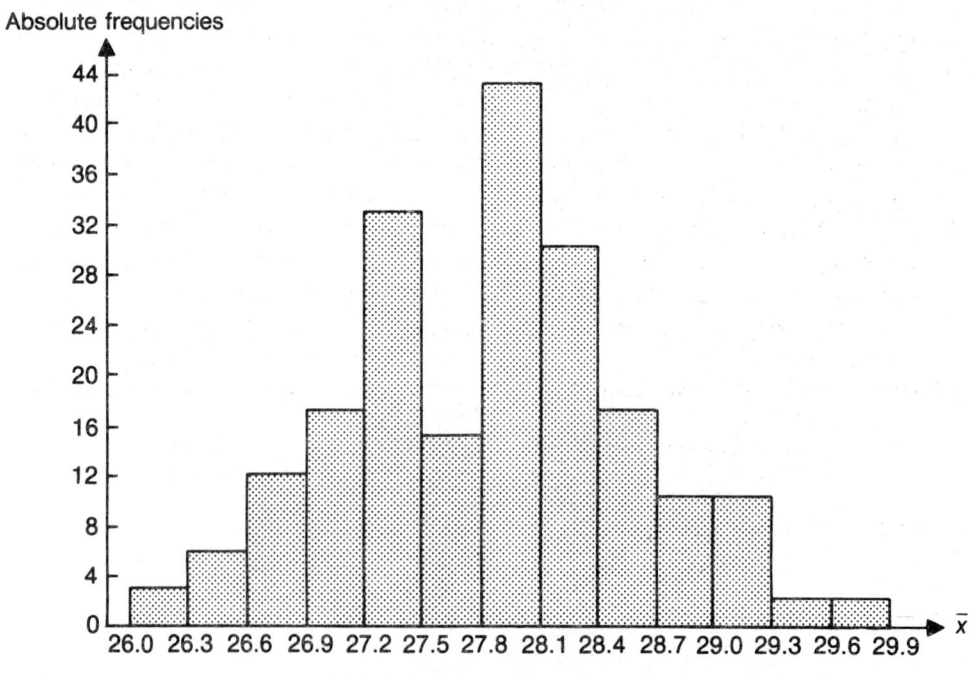

FIGURE C2.3   Frequency distribution and histogram for sample standard deviation from 200 samples of n = 15 ages

Values of s	Absolute frequency	Relative frequency
1.20–1.49	5	0.025
1.50–1.79	10	0.050
1.80–2.09	16	0.080
2.10–2.39	28	0.140
2.40–2.69	37	0.185
2.70–2.99	24	0.120
3.00–3.29	26	0.130
3.30–3.59	23	0.115
3.60–3.89	18	0.090
3.90–4.19	6	0.030
4.20–4.49	5	0.025
4.50–4.79	2	0.010
	200	1.000

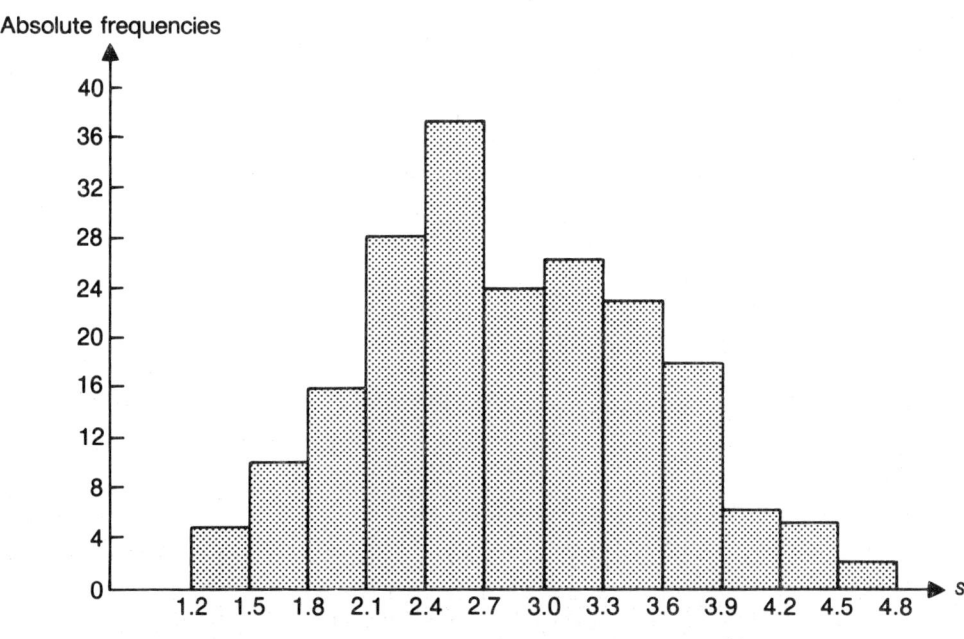

FIGURE C2.4   Frequency distribution and histogram from GSAMP for sample means from 2501 samples of n = 15 ages

Classes of elements		Absolute frequency	Abs.cumulat. frequency	Relative frequency	Rel.cumulat. frequency
25.400000	25.799999	2	2	0.001	0.001
25.800000	26.199999	18	20	0.007	0.008
26.200000	26.599999	62	82	0.025	0.033
26.600000	26.999999	206	288	0.082	0.115
27.000000	27.399999	341	629	0.136	0.251
27.400000	27.799999	482	1111	0.193	0.444
27.800000	28.199999	515	1626	0.206	0.650
28.200000	28.599999	418	2044	0.167	0.817
28.600000	28.999999	260	2304	0.104	0.921
29.000000	29.399999	138	2442	0.055	0.976
29.400000	29.799999	42	2484	0.017	0.993
29.800000	30.199999	11	2495	0.004	0.998
30.200000	30.599999	6	2501	0.002	1.000
		2501	2501	1.000	1.000

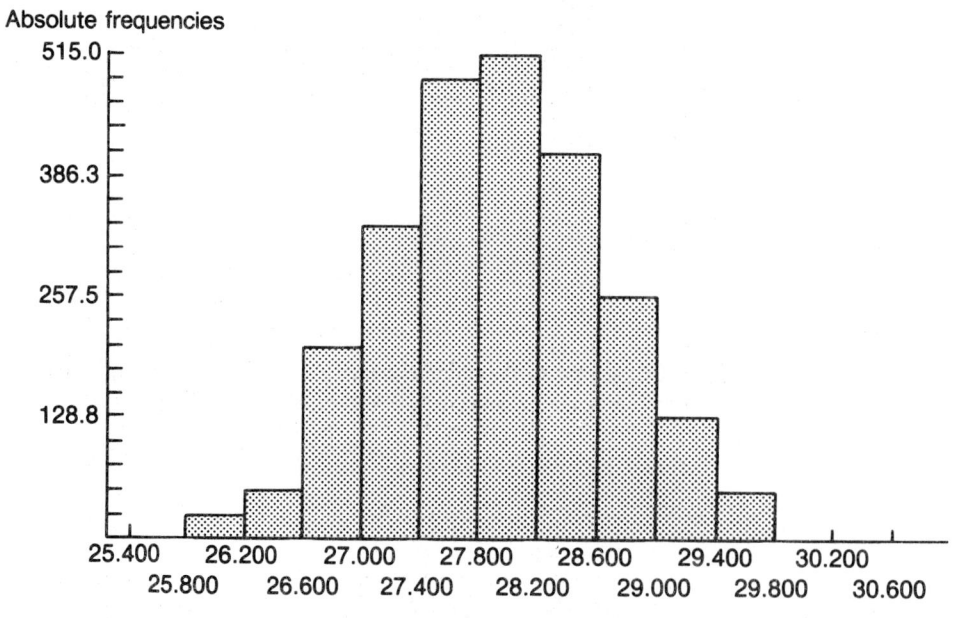

EXERCISES

C2.1.

In Exercise C1.3, use GSAMP to take 100 samples of 20 scores each and generate a frequency distribution and histogram of the 100 sample means.

C2.2.

In Exercise C1.1, use GSAMP to take 50 samples of 10 families each and display a frequency distribution and histogram of the 50 sample means.

C2.3.

Take the 50 sample standard deviations from the 50 samples found in Exercise C2.2, enter these 50 values in ISP, and use PFREQ and PHIST to obtain a frequency distribution and histogram of the 50 sample standard deviations.

C3

The Sampling Distribution of the Mean

In this chapter we focus on the sampling distribution of the sample mean. Certain properties of this sampling distribution are helpful when we use $\bar{x}$ to make inferences about the population mean. First, for a simple random sample of any size n taken from a population with mean μ, the mean of the sampling distribution of $\bar{x}$ equals μ:

$$\mu_{\bar{x}} = \mu \quad . \tag{C3-1}$$

Second, for a simple random sample (with replacement) of size n taken from a population with variance σ^2, the variance of the sampling distribution of $\bar{x}$ equals σ^2/n:

$$\sigma_{\bar{x}}^2 = \sigma^2/n \quad . \tag{C3-2}$$

The standard deviation of the sample mean is given a special name, the standard error of the mean:

$$\sigma_{\bar{x}} = \sigma/\sqrt{n} \quad . \tag{C3-3}$$

The standard error of the mean tells us exactly how the variability in sample means due to sampling fluctuations is reduced as the sample size increases. When we sample without replacement, the formula for the standard error must be adjusted slightly by multiplying by a term known as the finite population correction (fpc):

$$\text{fpc} = \sqrt{\frac{N - n}{N - 1}} \quad . \tag{C3-4}$$

When the population size N is large relative to n, the fpc is very close to 1, and any differences between sampling without replacement and sampling with replacement are very slight.

ISP COMMANDS

As indicated in the preceding chapters, GSAMP can be used to sample from any data set or from certain theoretical distributions. In GSAMP, standard errors are displayed. Some of these standard errors are estimates of $\sigma_{\bar{x}}$ because the sample standard deviation s is used to estimate σ in the formula for the standard error.

The ISP command STERR can be used to compute standard errors, with or without the finite population correction.

SOLVED EXAMPLE

Consider an achievement test that has been taken by several hundred thousand students. The scores on this test are normally distributed with a mean score of $\mu = 500$ and a standard deviation of $\sigma = 100$. If we take a random sample of n students' scores, we know from (C3–1) and (C3–3) that the sampling distribution of the sample mean $\bar{x}$ has mean

$$\mu_{\bar{x}} = \mu = 500$$

and standard error

$$\sigma_{\bar{x}} = \sigma/\sqrt{n} = 100/\sqrt{n} \ .$$

The fpc is ignored because the sample sizes we will consider in this example are miniscule relative to the population size.

Using GSAMP, we took repeated samples of size $n = 10$, $n = 25$, $n = 50$, $n = 100$, and $n = 500$ from the population of scores, and the results are given in Tables C3.1 through C3.5. First, consider the case of $n = 10$. Taking a single sample of $n = 10$ amounts to randomly selecting 10 students and recording their scores on the achievement test. The sample mean is just the average of their 10 scores. If we took another sample of 10 students, the individual scores and their average would most likely be different from the values found for the first sample. The sample means for 20 samples of size $n = 10$ each are shown in Table C3.1. These 20 values of $\bar{x}$ illustrate the variability of $\bar{x}$ due to sampling fluctuations. Although only 20 values of $\bar{x}$ are given in Table C3.1 to conserve space, GSAMP generated 2500 samples and found $\bar{x}$ for each of these samples. The overall mean and standard deviation of the 2500 sample means are 499.838 and 31.808, as given in Table C3.4 as part of the GSAMP printout. Using (C3–1) and (C3–3) with $n = 10$, we find $\mu_{\bar{x}} = 500$ and $\sigma_{\bar{x}} = 100/\sqrt{10} = 31.62$, which agree closely with the results of the repeated sampling via GSAMP.

Similar results for $n = 25$, $n = 50$, $n = 100$, and $n = 500$ are given in Tables C3.2 to C3.5. A total of 2500 samples of $n = 25$, 1000 samples of $n = 50$, 1000 samples of $n = 100$, and 500 samples of $n = 500$ were generated. Regardless of sample size,

TABLE C3.1   Some output from GSAMP for repeated samples of $n = 10$ achievement test scores

Sample	Sample Mean
1	445.6
2	517.7
3	522.4
4	489.9
5	482.2
6	513.4
7	460.5
8	446.6
9	497.8
10	465.8
11	441.5
12	481.8
13	413.2
14	530.1
15	537.9
16	444.8
17	478.2
18	498.0
19	511.0
20	514.0

Do you want more printing ? N
Overall mean of the 2500 sample means: 499.83801
Standard deviation: 31.807920

$\mu_{\bar{x}} = 500$. For the repeated sampling, the average of the sample means is 499.873 when $n = 25$, 499.689 when $n = 50$, 500.139 when $n = 100$, and 500.081 when $n = 500$. The standard error of the mean is, from (C3–3),

$$\sigma_{\bar{x}} = 100/\sqrt{25} = 20.00 \quad \text{when } n = 25 \quad,$$

$$\sigma_{\bar{x}} = 100/\sqrt{50} = 14.14 \quad \text{when } n = 50 \quad,$$

$$\sigma_{\bar{x}} = 100/\sqrt{100} = 10.00 \quad \text{when } n = 100 \quad,$$

$$\text{and } \sigma_{\bar{x}} = 100/\sqrt{500} = 4.47 \quad \text{when } n = 500 \quad.$$

From the repeated sampling, the standard deviation of the $\bar{x}$-values is

$$19.957 \quad \text{when } n = 25 \quad,$$

$$14.035 \quad \text{when } n = 50 \quad,$$

$$10.049 \quad \text{when } n = 100 \quad,$$

$$\text{and } 4.534 \quad \text{when } n = 500 \quad.$$

TABLE C3.2   Some output from GSAMP for repeated
samples of n = 25 achievement test scores

Sample	Sample Mean
1	525.1
2	510.6
3	500.6
4	520.6
5	502.7
6	548.9
7	501.9
8	529.4
9	494.3
10	459.4
11	507.8
12	502.8
13	521.3
14	474.1
15	511.5
16	516.0
17	482.7
18	521.0
19	521.5
20	540.5

Do you want more printing ? N
Overall mean of the 2500 sample means: 499.87344
Standard deviation: 19.957443

The variability of $\bar{x}$ in repeated samples decreases just as predicted by (C3–3) when n gets bigger. You can also get some idea of this reduced variability by looking at the 20 sample means given for each sample size in Tables C3.1 to C3.5. For the smaller samples of n = 10 and n = 25, values of $\bar{x}$ as low as 413.2 and as high as 548.9 are included in these brief lists. When n = 100, on the other hand, the lowest of the 20 $\bar{x}$-values is 471.3 and the highest is 514.2; and the 20 $\bar{x}$-values range from 492.4 to 508.5 when n = 500.

TABLE C3.3   Some output from GSAMP for repeated samples of $n = 50$ achievement test scores

Sample	Sample Mean
1	522.0
2	508.9
3	501.2
4	512.5
5	477.5
6	534.0
7	523.0
8	484.3
9	501.4
10	476.5
11	489.8
12	498.9
13	492.8
14	492.8
15	494.5
16	503.8
17	511.2
18	502.3
19	493.0
20	514.6

Do you want more printing ? N
Overall mean of the 1000 sample means: 499.68887
Standard deviation: 14.034533

TABLE C3.4 Some output from GSAMP for repeated samples of $n = 100$ achievement test scores

Sample	Sample Mean
1	512.8
2	509.2
3	498.8
4	484.1
5	498.5
6	491.7
7	504.8
8	505.9
9	490.5
10	491.8
11	494.5
12	505.8
13	479.9
14	471.3
15	514.2
16	506.5
17	505.8
18	498.0
19	505.9
20	488.7

Do you want more printing ? N
Overall mean of the 1000 sample means: 500.13928
Standard deviation: 10.049293

TABLE C3.5 Some output from GSAMP for repeated samples of $n = 500$ **achievement test scores**

Sample	Sample Mean
1	498.8
2	501.0
3	506.2
4	504.9
5	497.6
6	499.0
7	508.5
8	495.0
9	492.7
10	499.5
11	503.9
12	503.0
13	498.9
14	492.4
15	498.0
16	500.1
17	500.5
18	502.6
19	498.9
20	494.0

Do you want more printing ? N
Overall mean of the 500 sample means: 500.08148
Standard deviation: 4.5336056

EXERCISES

C3.1.

In Exercise C1.3, use GSAMP to take 100 samples of 20 scores each. Find the average of the 100 sample means and compare this with the population mean of 500. Find the standard deviation of the 100 sample means and compare this with the standard error derived from (C3–3), where the population standard deviation σ is 100.

C3.2.

In Exercise C1.2, use GSAMP to take 100 samples of size 10 with replacement and find the mean and standard deviation of the 100 sample means. Then take 100 samples of size 10 without replacement and find the mean and standard deviation of the 100 sample means. Compare the results for sampling with replacement with the results for sampling without replacement, and comment on any differences.

C3.3.

A sample of 10 students is taken, and their ages are 25, 19, 22, 22, 24, 20, 21, 26, 21, and 22. Enter these ages in ISP and use STERR to estimate the standard error of the mean.

C4
The Central Limit Theorem

The mean and standard error of the sampling distribution of the mean give us some idea about how much the sample mean is likely to differ from the population mean, but they do not provide enough information to enable us to determine exactly how likely a specific difference is. Fortunately, as long as the sample size is not too small (usually $n > 30$ is taken as a rule of thumb), an important theoretical result called the Central Limit Theorem comes to our aid:

> If the sample size n is large, the sampling distribution of the sample mean is approximately a normal distribution. The larger the sample, the better the approximation.

Note that no restriction is placed on the form of the population distribution. If the population itself follows a normal curve, then the sampling distribution of $\bar{x}$ is normal for *all n*. For non-normal populations, the sampling distribution of the mean will not generally be normal for small sample sizes.

When the sampling distribution of the mean is normal (either because the population is normal or because n is large enough for the Central Limit Theorem to apply), we can calculate probabilities concerning $\bar{x}$ from the normal distribution. Using (C3–1) and (C3–3), we have

$$z = \frac{\bar{x} - \mu_{\bar{x}}}{\sigma_{\bar{x}}} = \frac{\bar{x} - \mu}{\sigma/\sqrt{n}} \quad . \tag{C4–1}$$

Note that we divide by $\sigma/\sqrt{n}$, not by σ. We are standardizing $\bar{x}$, and the standard deviation of $\bar{x}$ is the standard error, $\sigma/\sqrt{n}$.

ISP COMMANDS

The command GSAMP can be used to generate a large number of repeated samples of a given sample size n from a data set or a theoretical distribution, and a histogram of the sample means can be displayed. This histogram provides an estimate of the sampling distribution of the mean. Generating such histograms for samples from different populations or distributions can provide a better understanding of the behavior of the sampling distribution of $\bar{x}$ and a better idea of how the Central Limit Theorem works.

As indicated in Chapter B9, the ISP command DNORM can be used to find probabilities for normally distributed variables. Thus, DNORM is helpful in finding probabilities for $\bar{x}$ when the Central Limit Theorem applies.

SOLVED EXAMPLE

A large company has 26,000 employees in various managerial positions classified as entry-level or middle management. Of these 26,000 employees,

> 10,000 have an annual salary of $20,000,
> 6,000 have an annual salary of $25,000,
> 4,000 have an annual salary of $30,000,
> 2,000 have an annual salary of $35,000,
> 2,000 have an annual salary of $40,000,
> 1,000 have an annual salary of $45,000,
> and   1,000 have an annual salary of $50,000.

FIGURE C4.1   Histogram from PHIST of managers' salaries (in thousands of dollars)

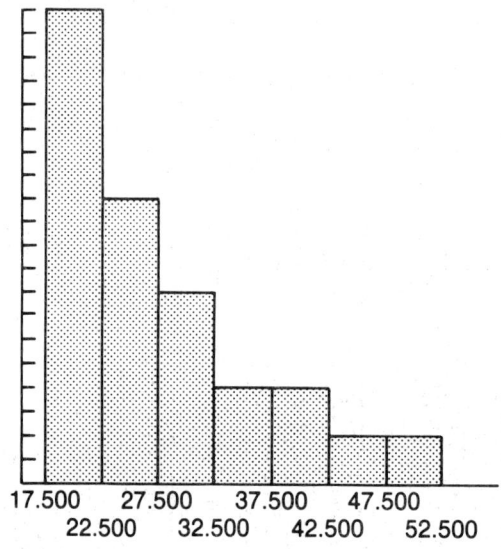

```
     17.500      27.500      37.500      47.500
         22.500      32.500      42.500      52.500
```

FIGURE C4.2 Histogram from GSAMP of 2500 sample means in repeated samples of n = 5 managers' salaries (in thousands of dollars)

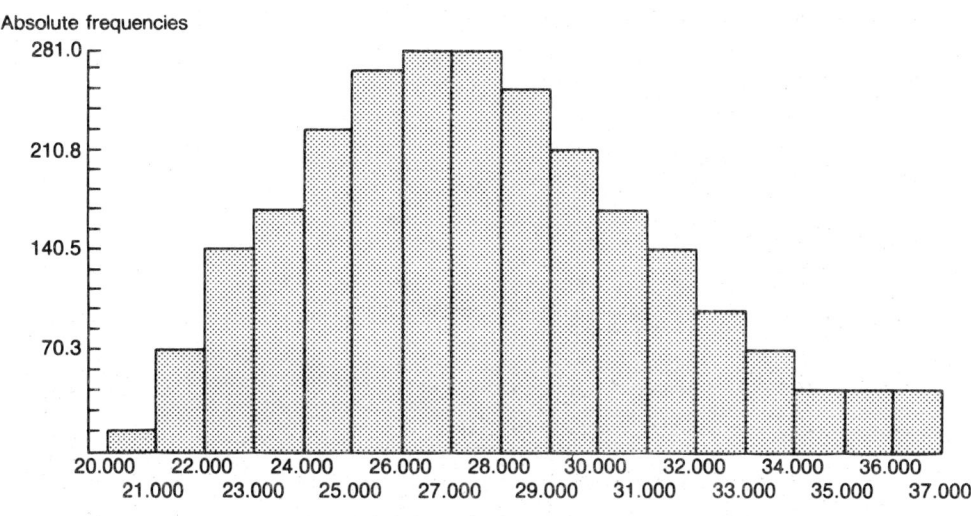

FIGURE C4.3 Histogram from GSAMP of 2500 sample means in repeated samples of n = 10 managers' salaries (in thousands of dollars)

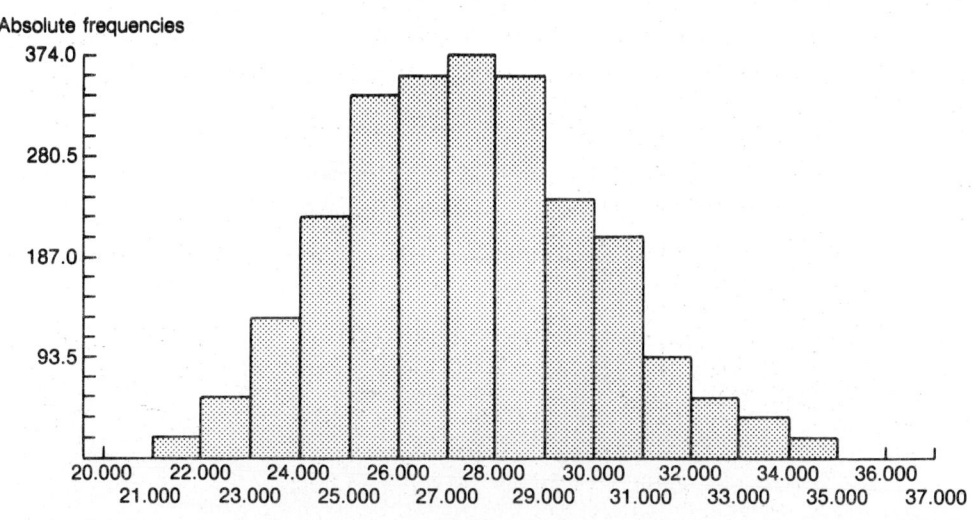

A histogram generated by PHIST for this population distribution of salaries is shown in Figure C4.1. Notice that the distribution is quite skewed, with a preponderance of salaries at the lower end of the range and a smaller number of salaries near the upper end of the range. The population is clearly *not* normally distributed.

The command GSAMP can be used to take repeated samples from this population of salaries. The sample mean $\bar{x}$ can be computed for each sample, and the histogram of $\bar{x}$-values in repeated samples of the same size n gives us an estimate of the sampling distribution of $\bar{x}$ for that value of n. Histograms based on 2500 samples each are given for $n = 5, 10, 20$, and 30 in Figures C4.2 to C4.5.

From Figure C4.2, it appears that the histogram of $\bar{x}$-values for samples of size $n = 5$ bears much more resemblance to a normal curve than does the population distribution. However, the histogram for samples of $n = 5$ is still somewhat skewed to the right, with a longer tail to the right than to the left. The population mean is $\mu = 27.5$ (in thousands of dollars), and there are quite a few $\bar{x}$-values in the 34–37 range (6.5–9.5 above the mean) but very few $\bar{x}$-values as low as 20–21 (6.5–7.5 below the mean).

When $n = 10$, the histogram of $\bar{x}$-values (Figure C4.3) is still skewed, but only slightly. The skewness is reduced considerably from the case of $n = 5$. Also, the distribution is less spread out (the standard error should be $\sigma/\sqrt{10}$ instead of $\sigma/\sqrt{5}$). The reduced skewness and decreased standard deviation of the distribution of $\bar{x}$-values continue as we move to $n = 20$ (Figure C4.4) and $n = 30$ (Figure C4.5). The distribution

FIGURE C4.4   Histogram from GSAMP of 2500 sample means in repeated samples of n = 20 managers' salaries (in thousands of dollars)

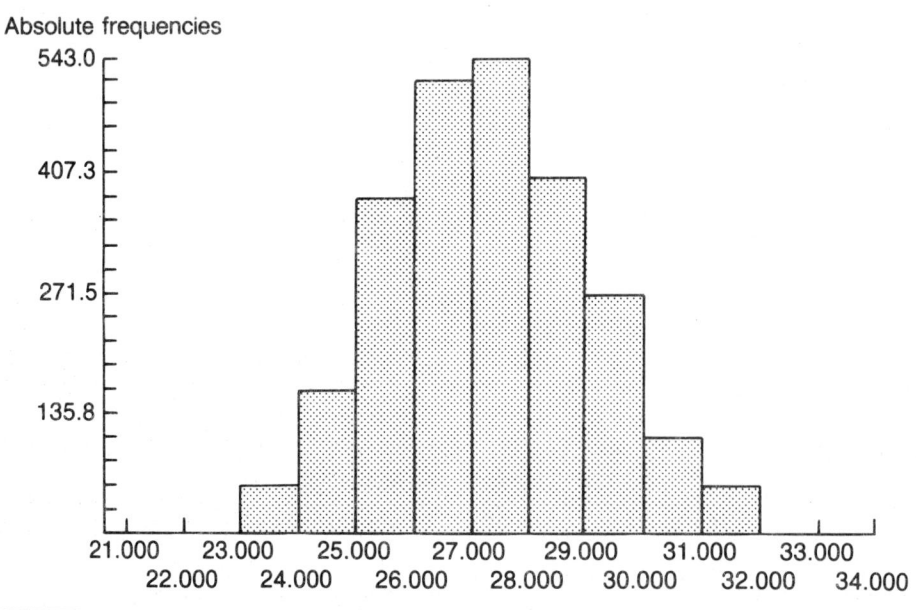

FIGURE C4.5   **Histogram from GSAMP of 2500 sample means in repeated samples of n = 30 managers' salaries (in thousands of dollars)**

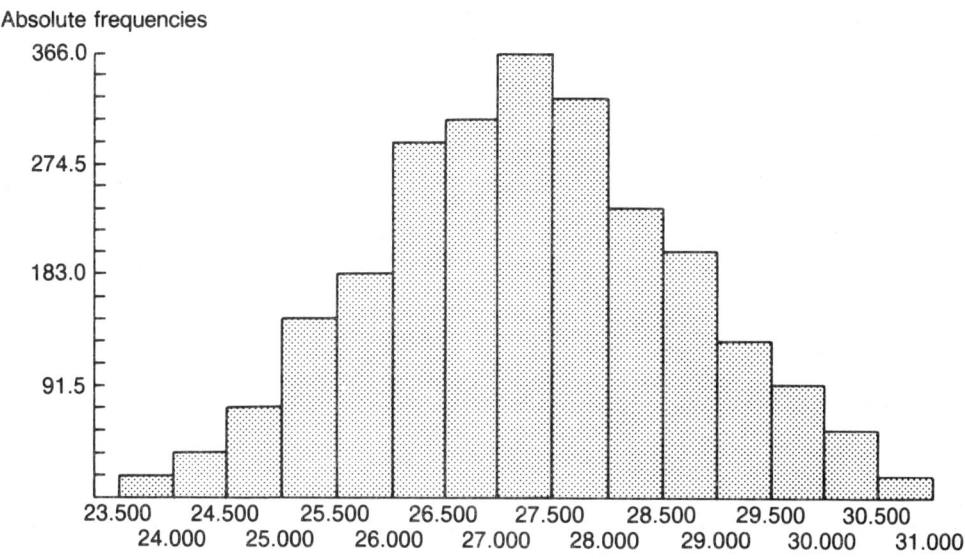

for $n = 30$ really does have a smaller standard deviation than the distribution for $n = 20$, although it may not appear so from a glance at Figures C4.4 and C4.5. The reason for the illusion is that, in order to give a better idea of how much the histogram for $n = 30$ looks like a normal curve, a smaller class interval (0.5, or $500) has been used in Figure C4.5 than in Figures C4.2 to C4.4 (which use class intervals of 1.0, or $1000).

This repeated sampling from an extremely non-normal population demonstrates how quickly the Central Limit Theorem works. Even for a sample size as small as $n = 10$, the resemblance of the histogram to the normal curve is not bad. For $n = 20$ and $n = 30$, the resemblance is quite good.

EXERCISES

C4.1.

Use GSAMP to take 1000 samples of size 2 from a uniform distribution with lower limit 0 and upper limit 10, and display a histogram of the 1000 sample means. Then repeat this procedure for samples of size 5, 10, and 30. Compare the four histograms obtained for the four different sample sizes, and comment on any differences.

C4.2.

Do Exercise C4.1 with a normal distribution with mean 5 and standard deviation 2 instead of a uniform distribution. Do the results seem to differ in nature from those found for the uniform distribution?

C4.3.

Use GSAMP to sample (with replacement) from a simple population consisting of eight 1s and two 0s. Take 2000 samples of size 5 and look at the histogram of the 2000 sample means; then repeat the process for $n = 10$ and $n = 30$. Compare the histograms.

C4.4.

Suppose that all you know about a population is that the mean is 70 and the standard deviation is 10. What can you say about the probability that the sample mean for a sample of size 4 from this population is greater than 72? What can you say about the probability that the sample mean for a sample of size 100 from this population is greater than 72?

C5

Using Sampling Distributions for Inferences

The primary purpose of taking a sample from a population is to use the data in the sample to make inferences about the population. In order to determine how accurate our inferences are, we need to know how large the sampling fluctuations are likely to be. Sampling distributions are helpful in statistical inference because they provide information about the variability of sample statistics. In estimation, we use a sample statistic to estimate a population parameter, and the sampling distribution tells us the probabilities that the estimate will differ from the true value of the parameter by certain amounts. In hypothesis testing, where some claim regarding the population parameter is being tested, we can say something about the likelihood of occurrence of certain values of the sample statistic if the claim is true. Information about sampling distributions can be very helpful in using sample data to make inferences about the population from which the sample was drawn.

ISP COMMANDS

The ISP commands useful in estimation and hypothesis testing will be introduced in the following chapters.

SOLVED EXAMPLE

In the salary example given in Chapter C4, a large company has 26,000 employees in various managerial positions classified as entry-level or middle management. In that chapter we assumed a particular population distribution of salaries for these managers and then took repeated samples from that population of salaries to illustrate the Central Limit Theorem. Of course, if we know exactly what the population distribution is like in an actual situation, there is no need to sample. The sampling in the salary example in Chapter C4 was for illustrative purposes only.

Suppose that we do *not* know the population distribution of salaries. We would like to obtain some information about this distribution, however. As a result, we take a random sample of $n = 50$ managers from the population and record their salaries. The 50 salaries are shown in Table C5.1. Be sure to notice that we are taking a *single* sample of size $n = 50$, not repeated samples of size $n = 50$.

The information from the sample can be used to make inferences about the entire population of 26,000 managers' salaries. For example, the sample mean $\bar{x} = 28.2$ and the sample standard deviation $s = 8.192$ can be calculated from the data by using SMEAN and SVAR, as shown in Figure C5.1. Thus, we estimate that the mean salary in the population is $28,200 and that the standard deviation is $8,192.

How accurate is our estimate of the mean? Our best guess about the standard error of the mean is

$$s/\sqrt{n} = 8.192/\sqrt{50} = 1.16 \quad ,$$

TABLE C5.1 Random sample of 50 managers' salaries (in thousands of dollars)

Observation	SAL	Observation	SAL
* 1*	30.000	* 26*	30.000
* 2*	40.000	* 27*	25.000
* 3*	20.000	* 28*	30.000
* 4*	30.000	* 29*	45.000
* 5*	30.000	* 30*	25.000
* 6*	40.000	* 31*	40.000
* 7*	40.000	* 32*	25.000
* 8*	25.000	* 33*	20.000
* 9*	30.000	* 34*	20.000
* 10*	25.000	* 35*	40.000
* 11*	35.000	* 36*	25.000
* 12*	20.000	* 37*	35.000
* 13*	20.000	* 38*	20.000
* 14*	50.000	* 39*	35.000
* 15*	35.000	* 40*	20.000
* 16*	20.000	* 41*	35.000
* 17*	20.000	* 42*	20.000
* 18*	25.000	* 43*	20.000
* 19*	30.000	* 44*	20.000
* 20*	25.000	* 45*	20.000
* 21*	25.000	* 46*	20.000
* 22*	35.000	* 47*	40.000
* 23*	20.000	* 48*	20.000
* 24*	25.000	* 49*	20.000
* 25*	25.000	* 50*	40.000

FIGURE C5.1   **Output from SMEAN, SVAR, and STERR for salary example**

⟨n⟩ SMEAN
$$***** \text{ Computes the Mean } *****$$

Which column(s) do you want to work on ? 1
Mean of your   50 numbers:

Column   1 (SAL   ) :   28.20000

⟨n⟩ SVAR
$$***** \text{ Computes the Variance and Standard Deviation } *****$$

Which column(s) do you want to work on ? 1
Here are the results for your   50 numbers:

	Variance	Standard Deviation
Column 1 (SAL) :	67.102043	8.1915836

These are the sample var. and stand. dev.—i.e. the divisor is $n-1$.

⟨n⟩ STERR
$$***** \text{ Computes the Standard Error } *****$$

Which column(s) do you want to work on ? 1
The standard error for your   50 numbers is:

Column   1 (SAL   ) :   1.1584649

as calculated by STERR in Figure C5.1. Because of the Central Limit Theorem, we can assume that the sampling distribution of the mean is normal. Therefore, the probability that an estimated mean salary from a sample of $n = 50$ is more than \$1000 higher than the true mean μ is

$$P(\bar{x} - \mu > 1) = P\left(z > \frac{1}{1.16}\right) = P(z > 0.86) = 0.195 \quad .$$

Similarly, the probability that an estimated mean salary is low by more than \$1000 is

$$P(\bar{x} - \mu < -1) = P\left(z < \frac{-1}{1.16}\right) = P(z < -0.86) = 0.195 \quad .$$

Thus, the probability that an estimated mean salary is in error by more than $1000 in either direction is

$$0.195 + 0.195 = 0.39 \quad .$$

The probability of being in error by more than $2000 in either direction is

$$P(\bar{x}-\mu>2) + P(\bar{x}-\mu < -2)$$

$$= P\left(z > \frac{2}{1.16}\right) + P\left(z < \frac{-2}{1.16}\right)$$

$$= P(z>1.72) + P(z<-1.72)$$

$$= 0.043 + 0.043 = 0.086 \quad .$$

Finally, the probability of being in error by more than $3000 in either direction is

$$P(\bar{x}-\mu>3) + P(\bar{x}-\mu<-3)$$

$$= P\left(z>\frac{3}{1.16}\right) + P\left(z<\frac{-3}{1.16}\right)$$

$$= P(z>2.59) + P(z<-2.59)$$

$$= 0.005 + 0.005 = 0.010 \quad .$$

These probabilities tell us how accurate $\bar{x}$ is likely to be as an estimate of μ. Since $\bar{x} = 28.2$ and the probability of being in error by more than $1000 is 0.39, we say that we are 61 percent confident that the population mean salary is in the interval from $27,200 to $29,200. Similarly, we are 91.4 percent confident that the mean salary is between $26,200 and $30,200; and we are 99 percent confident that the mean salary is between $25,200 and $31,200.

EXERCISES

C5.1.
   A sample of 64 students is to be taken from a very large population of students at a state university, and the age of each student in the sample will be recorded. If the standard deviation of age in the population is 4 years, what is the probability that the

sample mean $\bar{x}$ will be at least 1 year higher than the population mean age μ? How would this probability change if the sample size were only 16 instead of 64?

C5.2.

If the sample mean is used to estimate the population mean in Exercise C4.4, what is the probability of being in error by more than 2 in either direction?

C5.3.

Someone has claimed that the mean income in a community (in thousands of dollars) is 25. We suspect that μ is really higher than 25. If μ were really 25, what would the probability of getting a sample mean greater than 28 be, assuming that $\sigma = 10$ and $n = 36$?

Summary of Part C

When we collect data from a portion of a population, we say that we are sampling from the population. We can then use the procedures of Part A to describe or summarize the data that have been collected. In inferential statistics, however, we attempt to go beyond the data at hand to make inferences about the entire population. To do this, we need to know how the sample has been selected and how large the sampling fluctuations are likely to be. The purpose of Part C has been to prepare you for Parts D, E, and F by introducing you to the notion of sampling and to the important concept of a sampling distribution, which tells us something about sampling fluctuations.

To avoid systematic biases in sampling, statisticians use simple random sampling or some other probability sampling plan. With such sampling plans, we can determine how much variability to expect among samples. Sampling distributions of sample statistics such as $\bar{x}$ and s^2 tell us how much these statistics are likely to vary due to sampling fluctuations. For example, the standard error of the mean, which is the standard deviation of the sampling distribution of $\bar{x}$, is a measure of the variability of $\bar{x}$. Furthermore, since sampling distributions involve probability, the probability concepts covered in Part B are useful in dealing with sampling distributions. In particular, the Central Limit Theorem enables us to use the normal distribution to make statements about how much $\bar{x}$ is likely to differ from the population mean μ. The concept of a sampling distribution and results such as the Central Limit Theorem are crucial elements in statistical inference, which is the subject of the remainder of this book.

VOCABULARY LIST

accuracy of estimate

Central Limit Theorem

estimate

estimation

finite population correction

hypothesis testing

inference

inferential statistics

population

probability sampling

random numbers

random sample

representative sample

sample

sample size

sampling

sampling distribution

sampling distribution of the mean

sampling fluctuations

sampling without replacement

sampling with replacement

simple random sampling

standard error

standard error of the mean

stratified random sampling

systematic bias

LIST OF SYMBOLS

N	–	population size
μ	–	the mean of the population
σ^2	–	the variance of the population
σ	–	the standard deviation of the population
n	–	sample size
$\bar{x}$	–	the sample mean
s^2	–	the sample variance
s	–	the sample standard deviation
$\mu_{\bar{x}}$	–	the mean of the sampling distribution of $\bar{x}$
$\sigma_{\bar{x}}^2$	–	the variance of the sampling distribution of $\bar{x}$
$\sigma_{\bar{x}}$	–	the standard error of the mean
fpc	–	finite population correction
z	–	a standardized value

FORMULAS

$$\mu_{\bar{x}} = \mu$$

$$\sigma_{\bar{x}}^2 = \frac{\sigma^2}{n}$$

$$\sigma_{\bar{x}}^2 = \frac{\sigma^2}{n}\left(\frac{N-n}{N-1}\right) \qquad \text{(sampling without replacement)}$$

$$\sigma_{\bar{x}} = \frac{\sigma}{\sqrt{n}}$$

$$\sigma_{\bar{x}} = \frac{\sigma}{\sqrt{n}}\sqrt{\frac{N-n}{N-1}} \qquad \text{(sampling without replacement)}$$

$$\text{fpc} = \sqrt{\frac{N-n}{N-1}}$$

$$z \quad = \frac{\bar{x} - \mu_{\bar{x}}}{\sigma_{\bar{x}}} = \frac{\bar{x} - \mu}{\sigma/\sqrt{n}}$$

RELEVANT ISP COMMANDS

The command of greatest interest for the purposes of Part C is GSAMP, which allows you to take a single sample or repeated samples from a data set or a theoretical distribution. By taking repeated samples, you can observe sampling fluctuations and display distributions of sample means.

Other G commands also illustrate random sampling and display frequency distributions of sample statistics in repeated sampling. The following commands from the G group are useful in this regard.

GCARD:	Simulates a deck of playing cards
GCOIN:	Simulates the tossing of coins
GDIE:	Simulates the throwing of dice
GLOTO:	Simulates the game of LOTTO
GMARB:	Simulates sampling marbles from an urn
GROUL:	Simulates the game of roulette
GTACK:	Simulates the tossing of a thumbtack

The S commands, such as SMEAN, SMEDI, SMODE, SRANG, SVAR, and SUMST can be used to compute summary statistics. In addition, STERR computes the standard error of the mean.

Two D commands are helpful in Part C. DUNIF can be used to generate random numbers that can be used in random sampling. DNORM can provide probabilities from the normal distribution for cases in which the Central Limit Theorem applies and the sampling distribution of the mean is approximately normal.

As in Parts A and B, commands from the C, T, and P groups of commands can be useful for handling arithmetic computations, applying formulas, and printing and plotting data or probability distributions.

Finally, the MS commands briefly explain the topics covered in Part C. The MS modules correspond to the five chapters in Part C:

MSC1:	Sampling
MSC2:	Sampling distributions
MSC3:	The sampling distribution of the mean
MSC4:	The Central Limit Theorem
MSC5:	Using sampling distributions for inferences

PART D
Estimation and Hypothesis Testing

Having set the stage by introducing probability in Part B and sampling and sampling distributions in Part C, we are now prepared to discuss methods of statistical inference. In inferential problems, we use the information from a sample to say something about the population from which the sample was drawn. For example, the sample mean tells us something about the population mean. Furthermore, the sampling distribution of the sample mean tells us how much the sample mean is likely to differ from the population mean because of sampling fluctuations. Probabilities from the sampling distribution indicate how accurate our inferences are or how likely our inferences are to be in error.

Different types of problems require different types of inferences. The main distinction that is traditionally made is between problems of estimation and problems of hypothesis testing. In estimation problems, a sample statistic is used as an estimate of a population parameter, and we determine how large the error of estimation is likely to be. In hypothesis testing, specific claims concerning a population parameter are tested, and we decide whether to accept or reject these claims, taking into account the probabilities of erroneous decisions.

In Part D we discuss various estimation and hypothesis testing situations. Chapters D1 and D2 concern the estimation of means and proportions, respectively, for large samples. An approach called Bayesian estimation is presented in Chapter D3 for means and proportions. Chapters D4 and D5 also involve means and proportions (again with large samples), but in terms of hypothesis testing instead of estimation. Estimation and hypothesis testing for means when the sample size is small are discussed in Chapter D6. Chapter D7 describes methods for making inferences about variances. Finally, Chapter D8 involves inferences about the overall shape of the population distribution, a problem known in statistics as goodness of fit.

D1
Estimation of the Mean for Large Samples

When a population mean μ is of interest, the sample mean $\bar{x}$ provides an estimate of μ. Moreover, from what we know about the sampling distribution of the mean, we can find bounds on the error of estimation. A bound on the error of estimation with probability $1 - \alpha$ is

$$E = z_{\alpha/2}\sigma_{\bar{x}} = z_{\alpha/2}\left(\frac{\sigma}{\sqrt{n}}\right) \quad , \qquad \text{(D1–1)}$$

where $z_{\alpha/2}$ represents the value of z that "cuts off" $\alpha/2$ of probability in the right tail of a standard normal distribution. The bound on the error of estimation is $z_{\alpha/2}$ standard errors. For example, if a probability of 0.95 is desired, then $\alpha = 0.05$ and $z_{\alpha/2} = z_{0.025} = 1.96$, implying a bound of 1.96 standard errors (see Figure D1.1). We can use the notion of an error bound to determine the sample size required to give a specified degree of accuracy. If we want a bound E on the error of estimation with probability $1 - \alpha$, then the sample size should be

$$n = \left(\frac{z_{\alpha/2}\,\sigma}{E}\right)^2 \quad . \qquad \text{(D1–2)}$$

Another way of expressing the information provided by a point estimate and a bound on the error of estimation is a confidence interval. A confidence interval for μ with a confidence level $1 - \alpha$ is of the form

$$\bar{x} \pm z_{\alpha/2}\frac{\sigma}{\sqrt{n}} \qquad \text{(D1–3)}$$

if n is large. Thus, the lower confidence limit is

$$\bar{x} - z_{\alpha/2}\frac{\sigma}{\sqrt{n}} \qquad \text{(D1–4)}$$

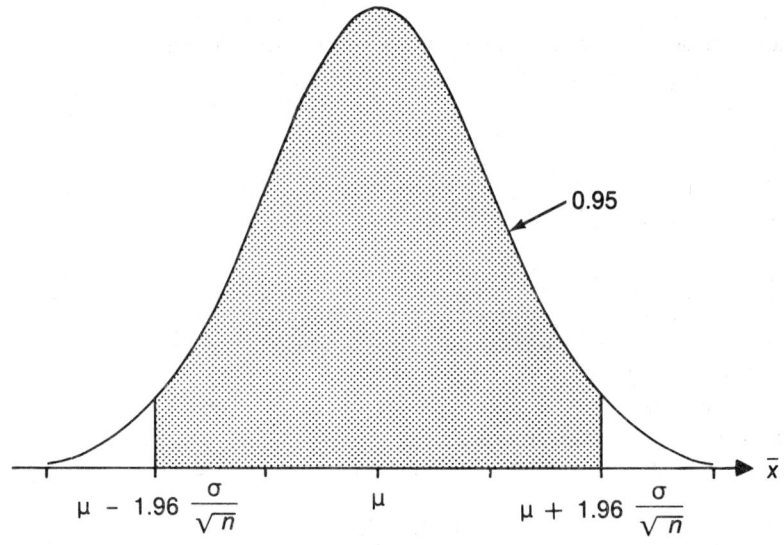

and the upper confidence limit is

$$\bar{x} + z_{\alpha/2} \frac{\sigma}{\sqrt{n}} \quad . \tag{D1-5}$$

From (D1-1), the expression being added to and subtracted from $\bar{x}$ is simply a bound on the error of estimation. In terms of a bound E, the confidence interval is

$$\bar{x} \pm E \quad , \tag{D1-6}$$

With confidence limits $\bar{x} - E$ and $\bar{x} + E$.

The above discussion assumes that the population standard deviation σ is known. In practice, σ (and therefore the standard error $\sigma/\sqrt{n}$) will not be known. Fortunately, we are dealing with large samples ($n > 30$) in this chapter. For such samples, the sample standard deviation s can safely be used in place of σ, giving us

$$s_{\bar{x}} = \frac{s}{\sqrt{n}} \quad , \tag{D1-7}$$

an estimate of $\sigma_{\bar{x}}$.

ISP COMMANDS

The command SEST provides error bounds and confidence intervals for μ. You need to enter the sample size, sample mean, and standard deviation (σ if known, s if σ is not known) and choose a confidence level. Of course, the sample mean and sample standard deviation can be computed by using the commands SMEAN and SVAR. If a level of confidence other than those available in SEST is desired, DNORM can be used to find the appropriate z-value.

SOLVED EXAMPLE

A hospital administrator is concerned about the number of days spent in the hospital by patients assigned to a particular surgical ward. Because virtually all patients on this ward are covered by insurance, there may be a tendency for the patients and their doctors to opt for a slightly longer stay in the hospital than is really necessary from a medical viewpoint. A random sample of 70 patients is chosen, and the number of days spent in the hospital by each of these patients is recorded. The data are displayed in Table D1.1.

The average number of days in the hospital for the patients in the sample is

$$\bar{x} = \frac{594}{70} = 8.49 \quad .$$

TABLE D1.1   Number of days spent in hospital for a sample of $n = 70$ patients

11	12	6	3
5	11	11	11
9	14	8	11
7	3	11	6
8	4	5	7
11	13	16	4
6	9	7	7
9	7	8	14
11	8	5	8
7	8	9	12
7	13	5	
6	12	5	
7	11	4	
7	8	6	
13	6	7	
13	9	4	
12	10	14	
13	7	7	
10	7	4	
8	5	12	

FIGURE D1.2   Output from SEST in hospital stay example

⟨n⟩ SEST
Which of the following parameters do you want to estimate:
1. The mean
2. The proportion
3. The variance or standard deviation
4. Differences between two means
5. Differences between two proportions
6. Ratios of two variances. (answer 1, 2, 3, 4, 5, or 6) ? 1
Sample size ? 70
Enter the sample mean : 8.49
Enter the standard deviation : 3.13
Do you want an error bound/confidence interval for :
1.   .90
2.   .95
3.   .99
Answer 1, 2, or 3 : 2
Population mean = 8.4900000 plus/minus 0.73324889
7.75675119 < population mean < 9.2232489

Thus, an estimate of the population mean μ is 8.49 days. There is considerable variability in length of stay, as might be expected because of different surgical procedures, the presence or absence of complicating factors, the age and condition of patients, and so on. The length of stay in the hospital ranges from 3 to 16 days in the sample, and the sample standard deviation, which can be computed with the command SVAR, is

$$s = 3.13 \quad .$$

An error bound with probability 0.95 and a 95 percent confidence interval are given in the output from SEST presented in Figure D1.2. The error bound of 0.73 is found from (D1–1) with s used in place of σ because n is large and σ is not known:

$$E = 1.96 \left(\frac{3.13}{\sqrt{70}} \right) = 0.73 \quad .$$

Thus, with probability 0.95, the error in using $\bar{x}$ to estimate μ in this example is at most 0.73 days. Expressing the same idea in a confidence interval, we get the following 95 percent confidence interval:

$$8.49 \pm 1.96 \left(\frac{3.13}{\sqrt{70}} \right) = 8.49 \pm 0.73 \quad ,$$

or

$$(7.76, 9.22) \quad ,$$

as shown in Figure D1.2. We are 95 percent confident that the mean length of stay in the hospital for patients in this ward is between 7.76 and 9.22 days.

Suppose the hospital administrator would like an error bound of 0.50 with probability 0.95. From (D1-2), using s in place of σ once again, the sample size required for this degree of accuracy is

$$n = \left[\frac{1.96\,(3.13)}{0.50}\right]^2 = 150.5 \quad.$$

To attain the desired degree of accuracy, the hospital administrator should take an additional random sample of 81 patients. Such a sample, combined with the sample of 70 observations in Table D1.1, would yield a total of n = 70 + 81 = 151 patients.

EXERCISES

D1.1.

A random sample of 40 cars of a particular model is tested for gasoline mileage. The sample mean is 35.24 miles per gallon, and the standard deviation is 4.18 miles per gallon. Find an error bound with probability 0.95 and a 95 percent confidence interval for μ, the mean gasoline mileage for this model, without using SEST. Then use SEST to obtain the error bound and confidence interval.

D1.2.

In order to estimate the mean weight of 18-year-old American males, a sample is to be taken. If we assume that the standard deviation of weight is 20 pounds and we want an error bound of 2 pounds with probability 0.99, how large a sample is needed?

D1.3.

If the data in Table A1.7 represent salaries for a random sample of 114 managers, use ISP to find a point estimate for μ, the mean salary of managers in the population from which the sample was drawn; a bound on the error of estimation with probability 0.90; and a 90 percent confidence interval for μ.

D2
Estimation of the Proportion for Large Samples

Just as a sample mean is a good estimate of a population mean, a sample proportion is a good estimate of a population proportion p. If the event of interest occurs x times in a sample of size n, the sample proportion is simply x/n. As long as the sample size is not too small, the sampling distribution of x/n is approximately normal with mean

$$\mu_{x/n} = p \tag{D2-1}$$

and standard error

$$\sigma_{x/n} = \sqrt{p(1-p)/n} \quad . \tag{D2-2}$$

Since we are dealing with large samples, we can substitute the estimate x/n for p in (D2–2) to get an estimated standard error. Thus, a bound on the error of estimation with probability $1 - \alpha$ when x/n is used to estimate p is

$$E = z_{\alpha/2} \sqrt{\left(\frac{x}{n}\right)\left(1 - \frac{x}{n}\right)\Big/ n} \quad . \tag{D2-3}$$

A confidence interval for p with confidence level $1 - \alpha$ is of the form

$$\frac{x}{n} \pm E \quad , \tag{D2-4}$$

which is, from, (D2–3),

$$\frac{x}{n} \pm z_{\alpha/2} \sqrt{\left(\frac{x}{n}\right)\left(1 - \frac{x}{n}\right)\Big/ n} \quad . \tag{D2-5}$$

When a particular degree of accuracy is desired, the required sample size is often found by a conservative approach that uses 0.5 for the proportion in the formula for the standard error of x/n. If E is the desired bound on the error with probability $1 - \alpha$, then (using the conservative approach) the sample size should be

$$n = \frac{z_{\alpha/2}^2}{4E^2} \quad .$$

(D2–6)

Alternatively, if a prior guess p^* is used instead of the conservative 0.50, we have

$$n = \frac{z_{\alpha/2}^2 \, p^* \, (1 - p^*)}{E^2} \quad .$$

(D2–7)

Unless the prior guess p^* is quite far from 0.50, the sample size found from (D2–7) will not differ substantially from that found via (D2–6).

ISP COMMANDS

The command SEST provides error bounds and confidence intervals not just for μ, but also for p and other parameters. You need to enter the sample size and the sample proportion and to choose a confidence level. For confidence levels other than 0.90, 0.95, and 0.99, the command DNORM can be used to find the appropriate z-value and the C commands can be used for the necessary calculations.

SOLVED EXAMPLE

The size of the audience for a television show is an important factor in finding advertisers to purchase time for commercials, in setting the rates for such time, and in determining whether or not the show should be discontinued. To estimate the proportion of families watching a particular show, a random sample of families is selected. Whether a family is watching the show or not is ascertained by having the family keep a record of their television watching, by attaching a device to the television that keeps such a record automatically, or by conducting a telephone interview. A random sample of $n = 150$ families reveals that $x = 33$ are watching the show.

Based on this sample, our point estimate of p, the proportion of families watching the show in the entire population, is

$$x/n = 33/150 = 0.22 \quad .$$

How much is x/n likely to vary because of sampling fluctuations? The estimated standard error of x/n is

$$\sqrt{\frac{x}{n}\left(1 - \frac{x}{n}\right)\Big/ n} = \sqrt{(0.22)\,(0.78)/150} = 0.0338 \quad.$$

An error bound with probability 0.95 and a 95 percent confidence interval for p are shown in the output from SEST presented in Figure D2.1. The error bound of 0.066 is computed from (D2–3):

$$E = 1.96\sqrt{(0.22)\,(0.78)/150} = 1.96\,(0.0338) = 0.066 \quad.$$

From (D2-4), then, a 95 percent confidence interval is

$$0.22 \pm 0.066 \quad,$$

or

$$(0.154,\ 0.286) \quad.$$

With 95 percent confidence, we estimate that the proportion of families watching the show is between 0.154 and 0.286. That is, we estimate that between 15.4 percent and 28.6 percent of the families in the population are watching the show.

FIGURE D2.1   Output from SEST in television show example

⟨n⟩ SEST

Which of the following parameters do you want to estimate :
1. The mean
2. The proportion
3. The variance or standard deviation
4. Differences beween two means
5. Differences between two proportions
6. Ratios of two variances. (answer 1, 2, 3, 4, 5, or 6) ? 2
Sample size ? 150
Enter the sample proportion : .22
Do you want an error bound/confidence interval for :
1. .90
2. .95
3. .99
Answer 1, 2, or 3 : 2
Population proportion = 0.220 plus/minus 0.066

$0.154 <$ population proportion < 0.286

The confidence interval of (0.154, 0.286) is quite wide. To improve the accuracy of the point estimate and to obtain a narrower 95 percent confidence interval, we could increase the sample size. For example, suppose that we observe $x = 66$ families watching the show in a sample of $n = 300$. Once again our point estimate of p is 0.22, since the sample proportion is $66/300 = 0.22$. But now, with probability 0.95, x/n will differ from p by at most

$$E = 1.96\sqrt{(0.22)(0.78)/300} = 0.047$$

in either direction. This is an improvement over the comparable error bound of 0.066 when $n = 150$. A 95 percent confidence interval for p when $n = 300$ and $x = 66$ is

$$0.22 \pm 0.047 \quad,$$

or

$$(0.173, 0.267) \quad.$$

To see how the accuracy of estimation changes when x/n is held constant and n is varied, error bounds and confidence intervals can be found for different values of n by using SEST or, equivalently, by using (D2–3) and (D2–4). For instance, error bounds and confidence intervals for $x/n = 0.22$ are given in Table D2.1 for samples of 150, 300, 1000, 5000, and 10,000 families. Large increases in n are required to obtain substantial reductions in the error bound and in the width of a 95 percent confidence interval.

Of course, because the estimated standard error depends on x/n as well as on n, the accuracy of estimation depends on x/n. If the sample of $n = 150$ families results in only $x = 15$ families watching the show, then $x/n = 15/150 = 0.10$. An error bound with probability 0.95 is then

$$1.96\sqrt{(0.10)(0.90)/150} = 0.048 \quad.$$

This is less than the error bound of 0.066 when $n = 150$ and $x/n = 0.22$. On the other

TABLE D2.1   Error bounds with probability 0.95 and 95 percent confidence intervals for p for different values of n when $x/n = 0.22$ in the television show example

n	Error bound	95% Confidence interval
150	0.066	(0.154, 0.286)
300	0.047	(0.173, 0.267)
1,000	0.026	(0.194, 0.246)
5,000	0.011	(0.209, 0.231)
10,000	0.008	(0.212, 0.228)

hand, if $x = 75$ out of $n = 150$ families are watching the show, then the error bound is *increased* to

$$1.96 \ \sqrt{(0.50) \ (0.50)/150} = 0.080 \quad .$$

For any n, the error bound is largest when $x/n = 0.50$.

Suppose that we would like to be 95 percent confident of estimating p within 0.01 in the television show example. Thus, with $1 - \alpha = 0.95$, the desired E is 0.01, which corresponds to an error of one percentage point if we think in terms of a percentage of the families in the population. To guarantee this degree of accuracy, we could take the conservative approach and use (D2–6):

$$n = \frac{(1.96)^2}{4(0.01)^2} = 9604 \quad .$$

If we feel quite strongly that less than 25 percent of the families watch the show, we might set $p^* = 0.25$ in (D2–7) and take a sample of

$$n = \frac{(1.96)^2(0.25)(0.75)}{(0.01)^2} = 7203$$

families. Either way (with $n = 9604$ or $n = 7203$, we see that a large sample is required to provide an error of at most 0.01, or one percentage point, with probability 0.95.

EXERCISES

D2.1.

In a random sample of 200 voters, 94 prefer Candidate A. Find a point estimate for the proportion of voters in the entire population who prefer Candidate A, and use SEST to determine an error bound with probability 0.95 and a 95 percent confidence interval for p.

D2.2.

An experiment is being designed to estimate the proportion of patients cured with a new drug. If we want an error bound of 0.03 with probability 0.90, determine the required sample size using the conservative approach. Then repeat the process using a prior guess of $p^* = 0.80$ for p.

D2.3.

Find a 99 percent confidence interval for p, the proportion of families in a particular community with more than one car, based on a sample of 90 families, 27 of which have more than one car.

D3

Bayesian Estimation

The procedures discussed in Chapters D1 and D2 for the estimation of means and proportions use information from a sample. We take a sample from a population and calculate estimates, error bounds, and confidence intervals from the data in the sample. This is called the sampling-theory approach, or classical approach, to estimation. The Bayesian approach to estimation enables us to combine sample information with prior information, which represents what we know before taking the sample. Estimates can then be based on this combined set of information (called posterior information because it represents what we know after, or posterior to, taking the sample).

We consider first the estimation of μ. Suppose that our prior information about μ can be represented by a normal distribution with mean m' and standard deviation σ'. The standard deviation σ' indicates that the accuracy of the prior information is equivalent to that from a sample of size

$$n' = \left(\frac{\sigma}{\sigma'} \right)^2 , \qquad\qquad (D3\text{--}1)$$

Where σ, as before, represents the population standard deviation. When Bayes' theorem is used to combine this prior information with the results of a sample, the resulting posterior distribution is a normal distribution with mean

$$m'' = \frac{n'm' + nm}{n' + n} \qquad\qquad (D3\text{--}2)$$

and standard deviation

$$\sigma'' = \frac{\sigma}{\sqrt{n' + n}} . \qquad\qquad (D3\text{--}3)$$

The posterior estimate of μ, m'', is a weighted average of the prior estimate m' and the sample mean $\bar{x}$, with the weights depending on the relative accuracy of m' and $\bar{x}$. A bound on the error of estimation in the posterior Bayesian estimate is

$$E = z_{\alpha/2}\sigma'' = z_{\alpha/2}\left(\frac{\sigma}{\sqrt{n'+n}}\right) \quad , \tag{D3-4}$$

and a Bayesian interval estimate for μ is

$$m'' \pm z_{\alpha/2}\left(\frac{\sigma}{\sqrt{n'+n}}\right) \quad . \tag{D3-5}$$

In a Bayesian analysis of a proportion, a convenient prior distribution to use is a beta distribution, which has a curve with formula

$$f(p) = \frac{(n'-1)!}{(x'-1)!(n'-x'-1)!}\, p^{x'-1}(1-p)^{n'-x'-1} \quad \text{for } 0 \le p \le 1 \quad . \tag{D3-6}$$

A beta distribution with parameters x' and n' can be thought of as conveying information roughly equivalent to a sample of size n' with the event of interest occurring x' times. The mean and standard deviation of the beta distribution given by (D3–6) are

$$m' = x'/n' \tag{D3-7}$$

and

$$\sigma' = \sqrt{m'(1-m')/(n'+1)} \quad . \tag{D3-8}$$

From (D3–7) and (D3–8), we can obtain formulas for n' and x':

$$n' = \frac{m'(1-m')}{(\sigma')^2} - 1 \tag{D3-9}$$

and

$$x' = m'n' \quad . \tag{D3-10}$$

With a beta prior distribution for p, the posterior distribution is also a beta distribution, but x' and n' are changed to

$$x'' = x' + x \tag{D3-11}$$

and

$$n'' = n' + \text{n}. \tag{D3--12}$$

The posterior estimate of p is simply

$$m'' = x''/n'', \tag{D3--13}$$

the mean of the posterior distribution. The standard deviation of the posterior distribution is

$$\sigma'' = \sqrt{m''(1 - m'')/(n'' + 1)} \ . \tag{D3--14}$$

Using a normal approximation, which is valid only for quite large values of n'', an error bound with probability $1 - \alpha$ when using m'' to estimate p is

$$E = z_{\alpha/2}\sigma'' = z_{\alpha/2} \sqrt{m''(1 - m'')/(n'' + 1)} \ , \tag{D3--15}$$

and the corresponding Bayesian interval estimate is

$$x''/n'' \pm z_{\alpha/2} \sqrt{m''(1 - m'')/(n'' + 1)} \ . \tag{D3--16}$$

ISP COMMANDS

The ISP command SBAYE can be used to find Bayesian point and interval estimates for means and proportions. The prior mean and standard deviation of μ of p must be entered, along with the appropiate sample statistics (n and $\bar{x}$ for estimation of μ, or n and x for estimation of p). The posterior distribution is provided by SBAYE, and Bayesian point and interval estimates from that posterior distribution are given.

SOLVED EXAMPLE

In the hospital stay example of Chapter D1, a hospital administrator is concerned about μ, the mean number of days spent in the hospital by patients assigned to a particular surgical ward. Suppose that the prior information about μ can be represented by a normal distribution with mean $m' = 9$ and standard deviation $\sigma' = 0.80$. A random sample of 70 patients yields the data presented in Table D1.1. From these 70 observations, we can compute the sample mean and standard deviation, which turn out to be $\bar{x} = 8.49$ and $s = 3.13$. Since n is large, we can use s in place of σ in all of our calculations.

To find the posterior distribution of μ and determine Bayesian point and interval estimates from this distribution, we can use SBAYE. Output from SBAYE for this example is presented in Figure D3.1. The posterior distribution for μ is a normal distribution with mean 8.582 and standard deviation 0.339. A posterior Bayesian estimate of μ is therefore 8.582, and a 95 percent interval estimate is (7.918, 9.246).

The calculations performed by SBAYE can be summarized briefly. First, from (D3–1), the accuracy of the prior distribution is equivalent to that provided by a sample of size

$$n' = \left(\frac{3.13}{0.80}\right)^2 = 15.31 \quad .$$

Next, the mean and standard deviation of the posterior distribution are, from (D3–2) and (D3–3),

$$m'' = \frac{15.31(9) + 70(8.49)}{15.31 + 70} = 8.582$$

FIGURE D3.1   Output from SBAYE for the estimation of μ in the hospital example

⟨n⟩ SBAYE

***** Revises Probabilities with Bayes' Theorem *****

Do you want to :
1. Revise probabilities for a single event
2. Revise probabilities for a number of events or classes

3. Revise a normal distribution for a mean
4. Revise a beta distribution for a proportion. (answer 1, 2, 3, or 4) ? 3

Enter the following information for the prior distribution of the population mean :
What is the mean of the prior distribution ? 9

What is the standard deviation of the prior distribution ? 0.80

Enter the following information for the sample :
What is the sample size ? 70

What is the sample mean ? 8.49

What is the standard deviation ? 3.13

The posterior distribution of the population mean is a normal distribution with mean   8.582 and standard deviation   0.339

A Bayesian estimate of the population mean is   8.582, and the probability is 0.95 that 7.918 < population mean <   9.246

FIGURE D3.2   **Output from SBAYE for the estimation of p in the hospital example**

⟨n⟩ SBAYE

***** Revises Probabilities with Bayes' Theorem *****

Do you want to :
1. Revise probabilities for a single event
2. Revise probabilities for a number of events or classes
3. Revise a normal distribution for a mean
4. Revise a beta distribution for a proportion. (answer 1, 2, 3, or 4) ? 4

Enter the following information for the prior distribution of the population proportion :
What is the mean of the prior distribution ? 0.90

What is the standard deviation of the prior distribution ? 0.06

A beta distribution with parameters   21.600 and   24.000 has a mean of   0.900 and a standard deviation of   0.060

Enter the following information for the sample :
What is the sample size ? 70

What is the number of successes x ? 59

The posterior distribution of the population proportion is a beta distribution with parameters   80.600 and   94.000. This distribution has a mean of   0.857 and a standard deviation of   0.036

A Bayesian estimate of the population proportion is   0.857 and the probability is   0.95 that 0.780 < population proportion <   0.920

and

$$\sigma'' = \frac{3.13}{\sqrt{15.31 + 70}} = 0.339 \quad .$$

The Bayesian point estimate is simply $m'' = 8.582$, and (D3–5) provides a 95 percent interval estimate:

$$8.582 \pm 1.96(0.339) = (7.918, 9.246) \quad .$$

From the posterior distribution, we are 95 percent sure that the mean time spent in the hospital by patients assigned to this surgical ward is between 7.918 and 9.246 days.

Suppose that the hospital administration would also like to know the proportion of patients p in this ward fully covered by hospitalization insurance. The prior mean and standard deviation of p are $m' = 0.90$ and $\sigma' = 0.06$, and in our random sample of

$n = 70$ patients we find $x = 59$ with full insurance. The output from SBAYE in Figure D3.2 shows that a Bayesian estimate of p is 0.857 and a 95 percent Bayesian interval is (0.78, 0.92).

When SBAYE is used to estimate a proportion, it is assumed that the prior distribution can be represented by a beta distribution. Using (D3–9) and (D3–10) with $m' = 0.90$ and $\sigma' = 0.06$, we get

$$n' = \frac{0.90(0.10)}{(0.06)^2} - 1 = 24$$

and

$$x' = 0.90(24) = 21.6 \quad .$$

This means that the prior information can be thought of as being equivalent to a sample of 24 patients, 21.6 of whom have full insurance. (Since this is not an actual sample, "equivalent" values for x' and n' need not be integers.)

The prior and sample information are combined via (D3–11) and (D3–12) to find $x'' = 21.6 + 59 = 80.6$ and $n'' = 24 + 70 = 94$, giving us a posterior mean of

$$m'' = \frac{80.6}{94} = 0.857$$

from (D3–13) and a posterior standard deviation from (D3–14) of

$$\sigma'' = \sqrt{\frac{0.857(0.143)}{95}} = 0.036 \quad .$$

The 95 percent interval provided by SBAYE is calculated directly from the beta distribution. The calculations are relatively complicated, however, and we will not present the details.

EXERCISES

D3.1.

In Exercise D1.1, assume that $\sigma = 4$ and that the prior distribution for μ is a normal distribution with mean 32 and standard deviation 10. Find a posterior Bayesian estimate of μ and a 95 percent Bayesian interval estimate for μ, using SBAYE.

D3.2.

In Exercise D1.3, suppose that the prior distribution for μ can be represented by a normal distribution with mean 38,000 and standard deviation 5000. Use SBAYE to find posterior Bayesian point and 95 percent interval estimates for μ.

D3.3.

In Exercise D2.1, the prior information about p can be represented in the form of a beta distribution with mean 0.40 and standard deviation 0.06. Use SBAYE to find posterior Bayesian point and interval estimates for p.

D3.4.

Before the sample is taken in Exercise D2.3, we have a beta prior distribution for p with mean 0.20 and standard deviation 0.09. Use SBAYE to find posterior point and 95 percent interval estimates for p.

D4

Hypothesis Testing for the Mean for Large Samples

In hypothesis testing, we begin with a claim or hypothesis that a parameter equals a particular value or falls in a particular interval of values. The claim, called the null hypothesis H_0, is then tested against an alternative hypothesis H_A. For hypothesis testing concerning μ, the null hypothesis generally will be of the form H_0: $\mu = \mu_0$, where μ_0 is the hypothesized value of μ. The alternative hypothesis typically will be either one-tailed to the left ($\mu < \mu_0$), one-tailed to the right ($\mu > \mu_0$), or two-tailed ($\mu \neq \mu_0$). The type of test appropriate in a given situation depends on the nature of the situation in terms of the alternative claim or hypothesis that is put forth.

When the sample is large enough to allow us to assume that the sampling distribution of $\bar{x}$ is approximately normal, we can use the standard normal distribution when working with the test statistic

$$z = \frac{\bar{x} - \mu_0}{\sigma/\sqrt{n}} \quad . \tag{D4--1}$$

As in Chapter D1, since the sample is large we can take $s/\sqrt{n}$ as an estimate of $\sigma/\sqrt{n}$ if the population standard deviation σ is not known.

The idea behind hypothesis testing is to decide whether $\bar{x}$ differs from μ enough (and in the appropriate direction for a one-tailed test) to cause us to reject H_0. In terms of z, we want to see if z differs from 0 by enough to suggest rejection of H_0. The z-value computed from (D4--1) is compared with critical z-values to see if it is in the rejection region. For a one-tailed test to the left, we reject the null hypothesis if z is less than or equal to some critical value; for a one-tailed test to the right, we reject if z is greater than or equal to some critical value. For a two-tailed test, we reject unless z is between two critical values.

The critical values depend on how large the probabilities of making errors are. Rejecting a null hypothesis when it is fact true is called a Type I error, and α is the probability of a Type I error. Accepting H_0 when it is false and H_A is true is called a Type II error, and it has probability β. A common rule of thumb is to set α equal to some small value such as 0.01 or 0.05 and determine a rejection region that provides this value of α. For example, if $\alpha = 0.01$ for a one-tailed test to the left, then we should reject if $z \leq -2.33$, since $P(z \leq -2.33) = 0.01$.

ISP COMMANDS

The command SHYP can be used to test hypotheses about μ and other parameters. For tests concerning a mean μ, the following information must be entered: the sample size, the value of μ_0 from the null hypothesis, the sample mean, the standard deviation (σ if known, s if σ is not known), the choice of α (0.01, 0.05, or 0.10), and the type of test (one-tailed to the left, one-tailed to the right, or two-tailed). SHYP provides the computed z-value, the critical value(s), and the decision (accept H_0 or reject H_0). For values of α other than 0.01, 0.05, and 0.10, the command DNORM can be used to find critical values.

SOLVED EXAMPLE

In the example from Chapter D1 involving the number of days spent in the hospital by patients assigned to a particular surgical ward, suppose that the average stay for patients in this type of ward is claimed to be 8 days. The hospital administrator wonders if the population mean for this hospital is currently greater than 8 days. The hypotheses of interest are

$$H_0: \mu = 8$$

and

$$H_A: \mu > 8 \quad .$$

The data from a random sample of $n = 70$ patients are presented in Table D1.1. The sample mean and sample standard deviation are

$$\bar{x} = 8.49$$

and

$$s = 3.13 \quad .$$

With this information, SHYP can be used for our one-tailed test to the right. The output from SHYP, using $\alpha = 0.05$, is displayed in Figure D4.1. The critical value of z for a one-tailed test to the right with $\alpha = 0.05$ is 1.64. The value of the z-statistic computed from the sample information, from (D4-1) with s used in place of σ because n is large and σ is not known, is

$$z = \frac{8.49 - 8}{3.13/\sqrt{70}} = 1.31 \quad .$$

FIGURE D4.1   Output from SHYP for one-tailed test to the right in hospital stay example

⟨n⟩ SHYP

***** Tests Hypotheses *****

Do you want to test a null hypothesis about :
1. The mean
2. The proportion
3. The variance or standard deviation
4. Differences between two means
5. Differences between two proportions
6. Ratios of two variances. (answer 1, 2, 3, 4, 5, or 6) ? 1
Sample size ? 70
Enter the population mean from the null hypothesis : 8
Enter the sample mean : 8.49
Enter the standard deviation : 3.13
Do you want an alpha (Type I error) of :
1. .10
2. .05
3. .01
Answer 1, 2, or 3 : 2
Do you want :
1. A two-tailed test
2. A one-tailed test to the right
3. A one-tailed test to the left (answer 1, 2, or 3) ? 2
The null hypothesis that the population mean is 8.0000000 can be accepted
The computed z-value is 1.310
The critical value is 1.640

The rejection region consists of $z \geq 1.64$, which corresponds to the upper 5 percent of the standard normal distribution. The computed z-value is less than 1.64 and is therefore not in the rejection region. In fact, the area under the normal curve to the right of $z = 1.31$ is

$$P(z \geq 1.31) = 0.095 \quad ,$$

as shown in Figure D4.2.

To illustrate a two-tailed test, we will use the same situation with a two-tailed alternative hypothesis. Instead of wondering whether the mean is greater than the claim of $\mu = 8$ days, the hospital administrator wonders whether the population mean for this hospital is different from the claim of $\mu = 8$ days in either direction. The hypotheses are now

$$H_0 : \mu = 8$$

FIGURE D4.2 Probability under a normal curve to the right of z = 1.31

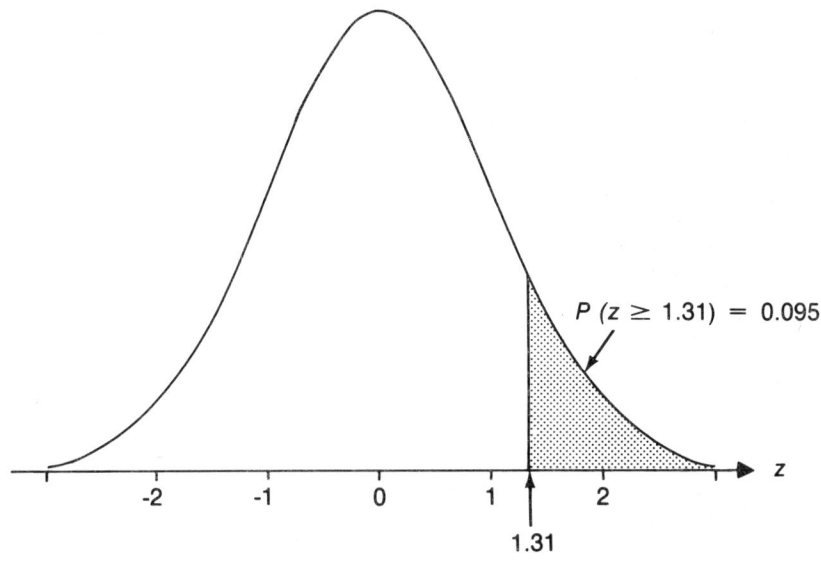

$P (z \geq 1.31) = 0.095$

1.31

and

$$H_A : \mu \neq 8 \quad .$$

The output from SHYP for this two-tailed test is shown in Figure D4.3. The rejection region is now $z \leq -1.96$ and $z \geq 1.96$. The computed z-value of 1.31 is clearly not in this rejection region.

FIGURE D4.3   Output from SHYP for two-tailed test in hospital stay example

⟨n⟩ SHYP

***** Tests Hypotheses *****

Do you want to test a null hypothesis about :
1. The mean
2. The proportion
3. The variance or standard deviation
4. Differences between two means
5. Differences between two proportions
6. Ratios of two variances. (answer 1, 2, 3, 4, 5, or 6) ? 1
Sample size ? 70
Enter the population mean from the null hypothesis : 8
Enter the sample mean : 8.49
Enter the standard deviation : 3.13
Do you want an alpha (Type I error) of :
1. .10
2. .05
3. .01
Answer 1, 2, or 3 : 2
Do you want
1. a two-tailed test
2. A one-tailed test to the right
3. A one-tailed test to the left (answer 1, 2, or 3) ? 1
The null hypothesis that the population mean is 8.0000000 can be accepted
The computed z-value is 1.310
The critical values are -1.960 and 1.960

EXERCISES

D4.1.

A company claims that students who have taken the company's speed typing course have a mean typing speed of 100 words per minute. The alternative hypothesis is that the company is too optimistic and that μ is in fact less than 100. A random sample of 100 students yields a sample mean of 94 words per minute with a standard deviation of 20 words per minute. Use SHYP to conduct a test with an alpha of 0.05.

D4.2.

A standardized test is designed to have a mean score of 500. Suppose that we want to test the claim that $\mu = 500$ against the alternative that μ is either higher or lower than 500. A random sample of 150 scores yields a sample mean score of 524.6 with a standard deviation of 112.5. If $\alpha = 0.10$, should we reject the claim that $\mu = 500$?

D4.3.

A builder claims that the mean monthly cost of heating one of his houses in the winter is $150. A random sample of size 60 provides a sample mean of $158 and a standard deviation of $30. Is this sufficient evidence to reject the builder's claim in favor of the alternative that μ is greater than $150, if $\alpha = 0.01$?

D5

Hypothesis Testing for the Proportion for Large Samples

Just as the estimation of proportions for large samples (Chapter D2) is similar to the estimation of means for large samples (Chapter D1), hypothesis testing for proportions is similar to hypothesis testing for means (Chapter D4). The basic notions of hypothesis testing (null and alternative hypotheses, one-tailed and two-tailed tests, Type I and Type II errors, rejection regions, critical values) apply to any statistical hypothesis testing, and for large samples (generally $n > 30$) the test statistic is a standard normal z-statistic for tests concerning p. If the null hypothesis is H_0: $p = p_0$, the test statistic is

$$z = \frac{(x/n) - p_0}{\sqrt{p_0(1 - p_0)/n}} \quad . \tag{D5-1}$$

Once the hypotheses are formulated and α is chosen, a rejection region can be determined by finding the critical z-value (or z-values in the case of a two-tailed test). If the z-value computed from the sample data is in the rejection region, we reject H_0; if not, we accept H_0. The only difference between tests for means and tests for proportions is that different formulas are needed to find the computed test statistic, or z-value.

ISP COMMANDS

The command SHYP can be used to test hypotheses about proportions as well as hypotheses about means and other parameters. For tests concerning a proportion p, the following information must be entered: the sample size, the value of p_0 from the null hypothesis, the sample proportion, the choice of α (0.01, 0.05, or 0.10), and the type of test (one-tailed to the left, one-tailed to the right, or two-tailed). SHYP provides the computed z-value, the critical value(s), and the decision (accept H_0 or reject H_0). For values of α other than 0.01, 0.05, and 0.10, the command DNORM can be used to find critical values.

SOLVED EXAMPLE

In the television show example in Chapter D2, a random sample of $n = 150$ families reveals that $x = 33$ are watching the show. The sample proportion is

$$x/n = 33/150 = 0.22 \quad .$$

Suppose that the producers of the show have claimed that $p = 0.28$ and have set advertising rates accordingly. Is the sample proportion low enough to cause us to reject the producers' claim?

We have the hypotheses

$$H_0 : p = 0.28$$

and

$$H_A : p < 0.28 \quad .$$

Let $\alpha = 0.01$, which means that the rejection region consists of $z \leqslant -2.33$ for this one-tailed test to the left. The computed value of the test statistic given by (D5–1) is

$$z = \frac{0.22 - 0.28}{\sqrt{(0.28)(0.72)/150}} = -1.64 \quad .$$

These calculations can be done by SHYP, as illustrated in Figure D5.1. Since the computed z-value of -1.64 is not in the rejection region, the sample proportion is apparently not low enough to cause us to reject the producers' claim at the $\alpha = 0.01$ level.

Our test tells us that the fact that x/n is 0.06 lower than the hypothesized value $(0.22 - 0.28 = -0.06)$ is not convincing evidence at the $\alpha = 0.01$ level that p is in fact less than the hypothesized value of 0.28. Of course, this is for a sample size of 150. What if we observed $x/n = 0.22$ for a larger sample? If $n = 400$, for instance, and $x = 88$, then $x/n = 0.22$ and

$$z = \frac{0.22 - 0.28}{\sqrt{(0.28)(0.72)/400}} = -2.67$$

as shown in the output from SHYP in Figure D5.2. Now z *is* in the rejection region. With a larger n, the difference of

$$0.22 - 0.28 = -0.06$$

is convincing evidence at the $\alpha = 0.01$ level than p is in fact less than the hypothesized value of 0.28.

FIGURE D5.1   **Output from SHYP for test in the television show example**

⟨n⟩ SHYP

***** Tests Hypotheses *****

Do you want to test a null hypothesis about :
1. The mean
2. The proportion
3. The variance or standard deviation
4. Differences between two means
5. Differences between two proportions
6. Ratios of two variances. (answer 1, 2, 3, 4, 5, or 6) ? 2
Sample size ? 150
Enter the population proportion from the null hypothesis : .28
Enter the sample proportion : .22
Do you want an alpha (Type I error) of :
1. .10
2. .05
3. .01
Answer 1, 2, or 3 : 3
Do you want :
1. A two-tailed test
2. A one-tailed test to the right
3. A one-tailed test to the left (answer 1, 2, or 3) ? 3
The null hypothesis that the population proportion is 0.280 can be accepted.
The computed z-value is -1.637
The critical value is -2.330

A sample proportion 0.06 below $p_0 = 0.28$ is not as unusual and surprising for a small sample as it is for a large sample. By way of analogy, you may not be surprised to get heads 7 times in 10 tosses of a coin you believe is fair. You would be very surprised, however, to get heads 700 times in 1000 tosses. Thus, a large sample is more sensitive to deviations from the hypothesized value than is a smaller sample. This is because the large sample provides greater accuracy. The sampling distributions of x/n if H_0 is true in the television show example are shown in Figure D5.3 for our two sample sizes, $n = 150$ and $n = 400$. When $n = 150$, the rejection region of $z \leqslant -2.33$ corresponds to

$$\frac{x}{n} \leqslant 0.28 - 2.33 \sqrt{(0.28)(0.72)/150} = 0.195 \quad .$$

When $n = 400$, the rejection region of $z \leqslant -2.33$ corresponds to

$$\frac{x}{n} \leqslant 0.28 - 2.33 \sqrt{(0.28)(0.72)/400} = 0.228 \quad .$$

FIGURE D5.2   Output from SHYP for test with $n = 400$ and $x/n = 0.22$ in the television show example

⟨n⟩ SHYP

***** Tests Hypotheses *****

Do you want to test a null hypothesis about :
1. The mean
2. The proportion
3. The variance or standard deviation
4. Differences between two means
5. Differences between two proportions
6. Ratios of two variances. (answer 1, 2, 3, 4, 5, or 6) ? 2
Sample size ? 400
Enter the population proportion from the null hypotheses : .28
Enter the sample proportion : .22
Do you want an alpha (Type I error) of :
1. .10
2. .05
3. .01
Answer 1, 2, or 3 : 3
Do you want :
1. A two-tailed test
2. A one-tailed test to the right
3. A one-tailed test to the left (answer 1, 2, or 3) ? 3
The sample evidence suggests that the null hypothesis that the proportion is 0.280 should be rejected.
The computed z-value is -2.673
The critical value is -2.330

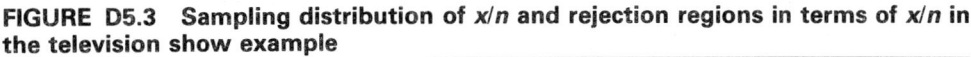

FIGURE D5.3 Sampling distribution of x/n and rejection regions in terms of x/n in the television show example

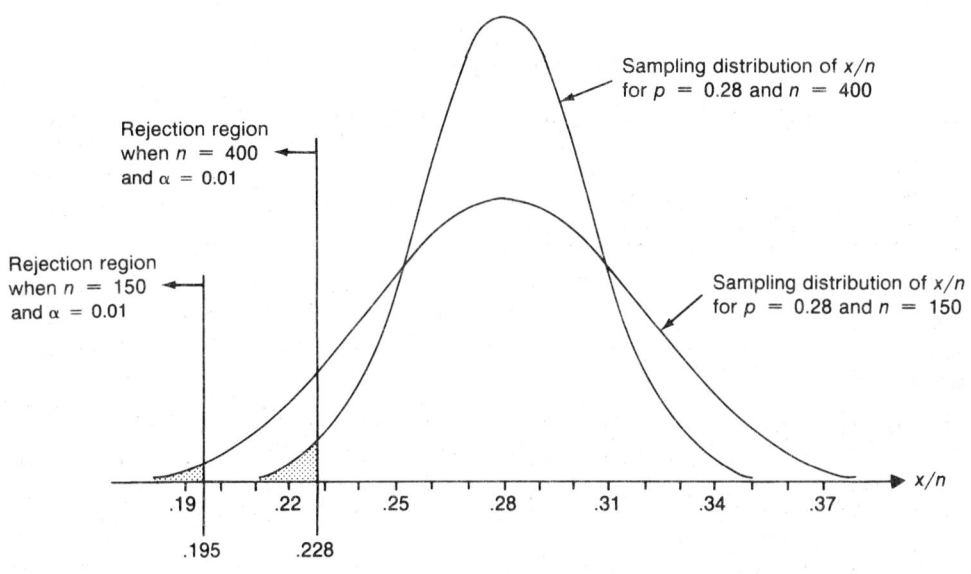

EXERCISES

D5.1.

A drug company claims that 80 percent of the people who use its new drug will be cured of minor asthma symptoms. Using a one-tailed test to the left with $\alpha = 0.05$, use SHYP to test this claim on the basis of a random sample of 230 patients to whom the drug was administered if 172 felt that the drug resulted in a cure.

D5.2.

A manufacturer claims that only 3 percent of its radios are defective. A random sample of 190 radios results in 8 defective radios and 182 non-defective radios. Is this strong enough evidence to cause us to reject the manufacturer's claim at the $\alpha = 0.01$ level in favor of the alternative that the percentage of defective radios is higher than 3 percent?

D5.3.

If a coin is tossed 500 times and comes up heads 238 times, should we reject the null hypothesis that the coin is fair in favor of a two-tailed alternative, assuming $\alpha = 0.05$?

D6
Inferences About the Mean for Small Samples

Inferences about the mean for small samples from normal populations require the use of the t distribution instead of the normal distribution. When n is small ($n \leq 30$) and the population distribution is assumed to be normal but σ is not known, then the statistic

$$t = \frac{\bar{x} - \mu}{s/\sqrt{n}} \qquad \text{(D6-1)}$$

has a t distribution with $n - 1$ degrees of freedom. A t distribution is symmetric, is centered at zero, and looks similar to but is more spread out than a standard normal distribution. As the number of degrees of freedom increases, the t distribution becomes more and more similar to a normal distribution (see Figure D6.1). Specific t values can be found in tables.

When $\bar{x}$ is used to estimate μ in a small-sample situation, an error bound with probability $1 - \alpha$ is

$$E = t_{\alpha/2}\left(\frac{s}{\sqrt{n}}\right) \quad , \qquad \text{(D6-2)}$$

where $t_{\alpha/2}$ is the t-value for a right-tail probability of $\alpha/2$. A confidence interval for μ with confidence level $1 - \alpha$ is

$$\bar{x} \pm t_{\alpha/2}\frac{s}{\sqrt{n}} \quad . \qquad \text{(D6-3)}$$

Finally, for testing hypotheses, the test statistic is

$$t = \frac{\bar{x} - \mu_0}{s/\sqrt{n}} \quad . \qquad \text{(D6-4)}$$

FIGURE D6.1   A standard normal curve and *t* curves with 3 and 9 degrees of freedom

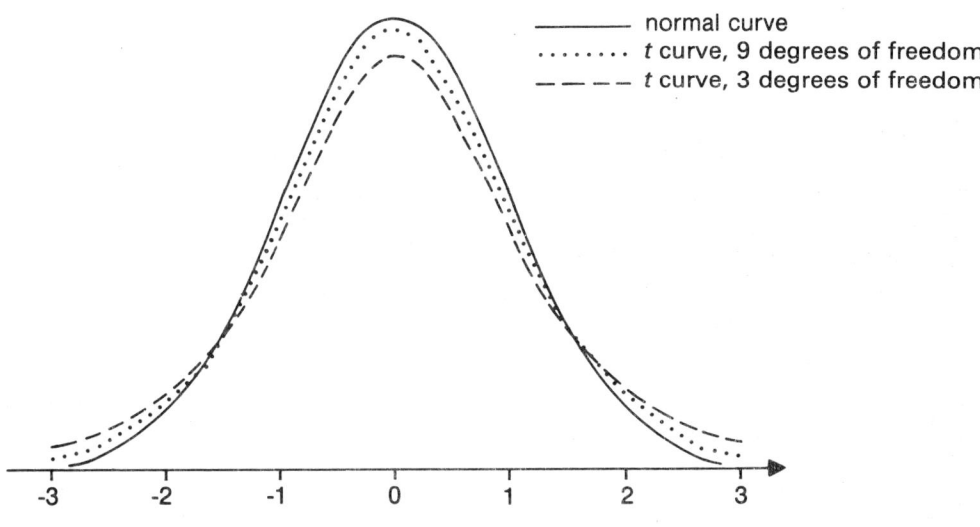

——— normal curve
········ *t* curve, 9 degrees of freedom
— — — — *t* curve, 3 degrees of freedom

Although the *t* distribution is used instead of the normal distribution, the basic methods for estimation and hypothesis testing remain unchanged.

ISP COMMANDS

The commands SEST and SHYP can be used for estimation and hypothesis testing about μ for small samples. When *n* is small, these commands automatically switch from *z*-values and *z*-statistics to *t*-values and *t*-statistics. SEST provides error bounds and confidence intervals for μ, and SHYP provides computed *t*-values, critical *t*-values, and decisions (accept or reject H_0). For levels of confidence and values of α not available in SEST and SHYP, the command DTDS can be used to find appropriate *t*-values. DTDS can provide a table of *t*-values, find a *t*-value for a given right-tail probability, and find a right-tail probability corresponding to a given *t*-value.

SOLVED EXAMPLE

A secretarial school claims that the average typing speed of its graduates is $\mu = 90$ words per minute. Some employers, however, have complained that recent graduates tend to be slower typists than a mean of 90 words per minute would suggest. A random sample of $n = 15$ recent graduates is chosen, and each is given a test to measure typing speed. The results are shown in Table D6.1.

TABLE D6.1   Typing speed, in words per minute, for a sample of $n = 15$ typists

76	80	100
78	75	89
81	86	91
90	102	67
84	66	69

The hypotheses of interest are

$$H_0 : \mu = 90$$

and

$$H_A : \mu < 90 \quad .$$

The sample mean is

$$\bar{x} = 82.27 \quad ,$$

which is less than 90. But is it far enough below 90 to cause us to reject H_0 at the $\alpha = 0.05$ level? The sample standard deviation is

$$s = 11.00 \quad ,$$

and the computed t-statistic for our test is, from (D6–4),

$$t = \frac{82.27 - 90}{11.00/\sqrt{15}} = -2.72 \quad .$$

This value is given in the output from SHYP displayed in Figure D6.2. The critical t-value for a one-tailed test to the left with $15 - 1 = 14$ degrees of freedom and $\alpha = 0.05$ is $t = -1.761$, which can be found from Table D6.2. Since $-2.72 < -1.761$, the hypothesis that $\mu = 90$ should be rejected.

If μ is not 90, what can we say about μ? The sample mean of 82.27 words per minute provides an estimate of μ. Furthermore, with probability 0.95 such an estimate is in error by at most

$$E = 2.145 \left(\frac{11.00}{\sqrt{15}} \right) = 6.09 \text{ words per minute.}$$

FIGURE D6.2   Output from SHYP for a one-tailed test to the left in typing speed example

$\langle n \rangle$ SHYP

***** Tests Hypotheses *****

Do you want to test a null hypothesis about :
1. The mean
2. The proportion
3. The variance or standard deviation
4. Differences between two means
5. Differences between two proportions
6. Ratios of two variances. (answer 1, 2, 3, 4, 5, or 6) ? 1

Sample size ? 15
Enter the population mean from the null hypothesis : 90
Enter the sample mean : 82.27
Enter the standard deviation : 11.00
Do you want an alpha (Type I error) of :
1. .10
2. .05
3. .01
Answer 1, 2, or 3 : 2
Do you want :
1. A two-tailed test
2. A one-tailed test to the right
3. A one-tailed test to the left (answer 1, 2, or 3) ? 3
The sample evidence suggests that the null hypothesis that the mean is 90.000000 should be rejected.
The computed t-value is -2.722
The critical value is -1.761

A 95 percent confidence interval for μ, as given in Figure D6.3, is

$$82.27 \pm 6.09 \quad ,$$

or

$$(76.18, 88.36) \quad .$$

We can state with confidence level 0.95 that the mean typing speed is between 76.18 and 88.36 words per minute.

TABLE D6.2   A table of t-values for certain specified right-tail probabilities

d.f.	.100	.050	.025	.010	.005
1	3.078	6.314	12.706	31.821	63.657
2	1.886	2.920	4.303	6.965	9.925
3	1.638	2.353	3.182	4.541	5.841
4	1.533	2.132	2.776	3.747	4.604
5	1.476	2.015	2.571	3.365	4.032
6	1.440	1.943	2.447	3.143	3.707
7	1.415	1.895	2.365	2.998	3.499
8	1.397	1.860	2.306	2.896	3.355
9	1.383	1.833	2.262	2.821	3.250
10	1.372	1.812	2.228	2.764	3.169
11	1.363	1.796	2.201	2.718	3.106
12	1.356	1.782	2.179	2.681	3.055
13	1.350	1.771	2.160	2.650	3.012
14	1.345	1.761	2.145	2.624	2.977
15	1.341	1.753	2.131	2.602	2.947
16	1.337	1.746	2.120	2.583	2.921
17	1.333	1.740	2.110	2.567	2.898
18	1.330	1.734	2.101	2.552	2.878
19	1.328	1.729	2.093	2.539	2.861
20	1.325	1.725	2.086	2.528	2.845
21	1.323	1.721	2.080	2.518	2.831
22	1.321	1.717	2.074	2.508	2.819
23	1.319	1.714	2.069	2.500	2.807
24	1.318	1.711	2.064	2.492	2.797
25	1.316	1.708	2.060	2.485	2.787
26	1.315	1.706	2.056	2.479	2.779
27	1.314	1.703	2.052	2.473	2.771
28	1.313	1.701	2.048	2.467	2.763
29	1.311	1.699	2.045	2.462	2.756

FIGURE D6.3   Output from SEST in typing speed example

⟨n⟩ SEST

Which of the following parameters do you want to estimate :
1. The mean
2. The proportion
3. The variance or standard deviation
4. Differences between two means
5. Differences between two proportions
6. Ratios of two variances. (answer 1, 2, 3, 4, 5, or 6) ? 1
Sample size ? 15
Enter the sample mean : 82.27
Enter the standard deviation : 11.00
Do you want an error bound/confidence interval for :
1. .90
2. .95
3. .99
Answer 1, 2 or 3 : 2
Population mean = 82.270000   plus/minus   6.0922027

76.177797   < population mean <   88.362203

EXERCISES

D6.1.
Using the data from Table A1.5, determine a point estimate of μ, the mean number of accidents reported per week at a busy intersection. Also, find a bound on the error of estimation with probability 0.95 and a 95 percent confidence interval for μ.

D6.2.
In the standard normal distribution, $z = 1.96$ cuts off a probability of 0.025 in the right tail of the distribution. What value of t cuts off this same probability when we have 5 degrees of freedom? Use DTDS to find the appropriate value of t and repeat the procedure with 10 and 15 degrees of freedom.

D6.3.
A health club claims that the mean pulse rate of individuals in its aerobics program is 60. A random sample of size 6 yields pulse rates of 71, 59, 64, 55, 69, and 62. Is this sufficient evidence, at the $\alpha = 0.01$ level, to conclude that the mean pulse rate is greater than 60?

D7
Inferences About the Variance

Just as the sample mean $\bar{x}$ provides a good estimate of the population mean μ, the sample variance s^2 provides a good estimate of the population variance σ^2. To say something about the accuracy of estimation, we need to consider the sampling distribution of s^2. First, the mean of the sampling distribution of s^2 is σ^2, implying that s^2 provides an unbiased estimate of σ^2. Second, if the population is approximately normally distributed, then s^2 can be related to a sampling distribution called the chi-square distribution, often written χ^2 distribution. In particular, the statistic

$$\chi^2 = \frac{(n-1)s^2}{\sigma^2} \qquad \text{(D7–1)}$$

has a chi-square distribution with $n - 1$ degrees of freedom. Some examples of chi-square distributions are given in Figure D7.1. A chi-square variable can never be negative, and the distribution is skewed to the right. The mean equals the number of degrees of freedom, and the variance is twice the number of degrees of freedom. Specific χ^2-values can be found in tables.

The general form for a confidence interval for σ^2 with confidence level $1 - \alpha$ is

$$\left(\frac{(n-1)s^2}{\chi^2_{\alpha/2}} \, , \, \frac{(n-1)s^2}{\chi^2_{1-(\alpha/2)}} \right) \, , \qquad \text{(D7–2)}$$

where $\chi^2_{\alpha/2}$ is the chi-square value cutting off probability $\alpha/2$ in the right tail of the distribution and $\chi^2_{1-(\alpha/2)}$ cuts off probability $1 - \frac{\alpha}{2}$ to the right. Note that the higher χ^2-value is used to compute the lower limit of the confidence interval and the lower χ^2-value is used in the calculation of the upper limit of the confidence interval. Note also that because of the asymmetry of the χ^2 distribution, the confidence interval is *not* of the form $s^2 \pm E$.

The test statistic for a test involving σ^2 is

$$\chi^2 = \frac{(n-1)s^2}{\sigma_0^2} \quad , \tag{D7-3}$$

where σ_0^2 is the value of σ^2 specified in the null hypothesis. In general, for inferences about a variance we simply use χ^2-values and compute χ^2-statistics instead of dealing with the normal and t distributions. When n is large ($n > 30$), a normal approximation with mean $n-1$ and standard deviation $\sqrt{2(n-1)}$ can be used to find χ^2-values.

ISP COMMANDS

The commands SEST and SHYP can be used for estimation and hypothesis testing concerning σ^2. SEST provides confidence intervals for the variance, and SHYP provides computed χ^2-values, critical χ^2-values, and decisions (accept or reject H_0). For levels of confidence and values of α not available in SEST and SHYP, the command DCHI can be used to find appropriate χ^2-values. DCHI can provide a table of χ^2-values, find a χ^2-value for a given right-tail probability, and find a right-tail probability corresponding to a given χ^2-value.

FIGURE D7.1   **Chi-square curves for various degrees of freedom**

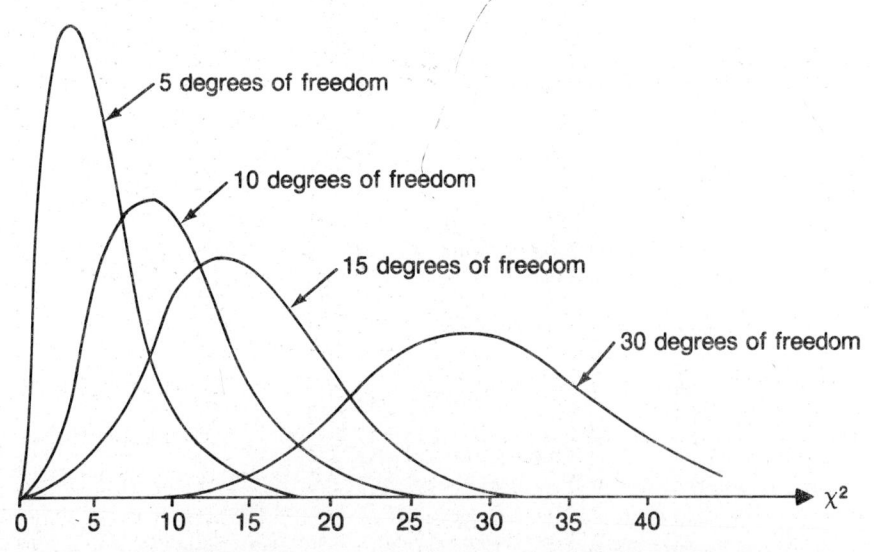

SOLVED EXAMPLE

A dairy would like to maintain a low level of variability in the butterfat content of its milk. If the butterfat content is supposed to be 2 percent, for example, then the actual percentage of butterfat in a carton of milk should not deviate much from this figure. A standard deviation of 0.10 percent is acceptable, but larger standard deviations are not. Since a standard deviation of 0.10 corresponds to a variance of $(0.10)^2 = 0.01$, the hypotheses of interest are

$$H_0 : \sigma^2 = 0.01$$

and

$$H_A : \sigma^2 > 0.01 \quad .$$

Table D7.1 presents the results of a random sample of $n = 20$ cartons of milk. The percentage of butterfat in these cartons ranges from 1.63 to 2.23, and the sample variance is

$$s^2 = 0.02237 \quad ,$$

as computed with SVAR and shown in Figure D7.2. The standard deviation is

$$s = 0.15 \text{ percent} \quad .$$

The χ^2-statistic for this example is, from (D7–3),

$$\chi^2 = \frac{(20 - 1)(0.02237)}{0.01} = 42.37 \quad ,$$

TABLE D7.1   Percentage of butterfat in $n = 20$ cartons of milk

1.83	2.00
2.22	2.07
1.98	2.02
1.88	2.05
1.95	2.12
1.78	1.91
2.03	2.06
2.23	2.15
1.63	1.89
1.84	1.91

FIGURE D7.2   Output from SVAR in butterfat content example

⟨n⟩ SVAR

***** Computes the Variance and Standard Deviation *****

Which column(s) do you want to work on ? 1
Here are the results for your 20 numbers :

		Variance	Standard deviation
Column 1 (PERBF)	:	0.22371955E-01	0.14957258

these are the sample variance and standard deviation—i.e., the divisor is n − 1.

FIGURE D7.3   Output from SHYP for one-tailed test to the right in butterfat content example

Do you want to test a null hypothesis about :
1. The mean
2. The proportion
3. The variance or standard deviation
4. Differences between two means
5. Differences between two proportions
6. Ratios of two variances. (answer 1, 2, 3, 4, 5, or 6) ? 3
Sample size ? 20
Enter the population variance from the null hypothesis : .01
Enter the sample variance : .02237
Do you want an alpha (Type I error) of :
1. .10
2. .05
3. .01
Answer 1, 2, or 3 : 2
Do you want :
1. A two-tailed test
2. A one-tailed test to the right
3. A one-tailed test to the left (answer 1, 2, or 3) ? 2
The sample evidence suggests that the null hypothesis that the variance is
0.0100000   should be rejected
The computed Chi-square value is   42.370
The critical value is 30.144

as shown in the output from SHYP in Figure D7.3. With $\alpha = 0.05$ and $20 - 1 = 19$ degrees of freedom, the critical value of χ^2 is 30.144, which can be found in Table D7.2. But the computed χ^2-value of 42.37 is greater than 30.144, implying that H_0 should be rejected in this one-tailed test to the right.

When we reject H_0 in this example, we conclude that σ^2 is indeed larger than 0.01. Our point estimate of σ^2 is the sample variance, 0.02237, and a confidence interval for σ^2 can be found from (D7–2). For a 95 percent confidence interval, $1 - \alpha = 0.95$.

TABLE D7.2   A table of chi-square values for certain right-tail probabilities

d.f.	.995	.990	.975	.950	.050	.025	.010	.005
1	0.000	0.000	0.001	0.004	3.841	5.024	6.635	7.879
2	0.010	0.020	0.051	0.103	5.991	7.378	9.210	10.597
3	0.072	0.115	0.216	0.352	7.815	9.348	11.345	12.838
4	0.207	0.297	0.484	0.711	9.488	11.143	13.277	14.860
5	0.412	0.554	0.831	1.145	11.070	12.832	15.086	16.750
6	0.676	0.872	1.237	1.635	12.592	14.449	16.812	18.548
7	0.989	1.239	1.690	2.167	14.067	16.013	18.475	20.278
8	1.344	1.646	2.180	2.733	15.507	17.535	20.090	21.955
9	1.735	2.088	2.700	3.325	16.919	19.023	21.666	23.589
10	2.156	2.558	3.247	3.940	18.307	20.483	23.209	25.188
11	2.603	3.053	3.816	4.575	19.675	21.920	24.725	26.757
12	3.074	3.571	4.404	5.226	21.026	23.337	26.217	28.300
13	3.565	4.107	5.009	5.892	22.362	24.736	27.688	29.819
14	4.075	4.660	5.629	6.571	23.685	26.119	29.141	31.319
15	4.601	5.229	6.262	7.261	24.996	27.488	30.578	32.801
16	5.142	5.812	6.908	7.962	26.296	28.845	32.000	34.267
17	5.697	6.408	7.564	8.672	27.587	30.191	33.409	35.718
18	6.265	7.015	8.231	9.390	28.869	31.526	34.805	37.156
19	6.844	7.633	8.907	10.117	30.144	32.852	36.191	38.582
20	7.434	8.260	9.591	10.851	31.410	34.170	37.566	39.997
21	8.034	8.897	10.283	11.591	32.671	35.479	38.932	41.401
22	8.643	9.542	10.982	12.338	33.924	36.781	40.289	42.796
23	9.260	10.196	11.689	13.091	35.172	38.076	41.638	44.181
24	9.886	10.856	12.401	13.848	36.415	39.364	42.980	45.558
25	10.520	11.524	13.120	14.611	37.652	40.646	44.314	46.928
26	11.160	12.198	13.844	15.379	38.885	41.923	45.642	48.290
27	11.808	12.879	14.573	16.151	40.113	43.194	46.963	49.645
28	12.461	13.565	15.308	16.928	41.337	44.461	48.278	50.993
29	13.121	14.256	16.047	17.708	42.557	45.722	49.588	52.336
30	13.787	14.953	16.791	18.493	43.773	46.979	50.892	53.672

Therefore, $\alpha = 0.05$ and $\alpha/2 = 0.025$. From a chi-square table in the row for 19 degrees of freedom,

$$\chi^2_{0.025} = 32.852$$

and

$$\chi^2_{0.975} = 8.907 \quad .$$

The 95 percent confidence interval for σ^2 is

$$\left(\frac{19(0.02237)}{32.852} , \frac{19(0.02237)}{8.907} \right) ,$$

or

$$(0.013 , 0.048) \quad .$$

This can be found from SEST, as illustrated in Figure D7.4. For a 95 percent confidence interval for the standard deviation σ, we just take square roots to convert from σ^2 to σ :

$$(\sqrt{0.013} , \sqrt{0.048}) \quad ,$$

or

$$(0.11 , 0.22) \quad .$$

We are 95 percent confident that the standard deviation of butterfat content is between 0.11 and 0.22 percent.

To get an idea of how the sample size affects the accuracy of estimation of the variance, we will consider a sample with a larger n of 50, but the same sample variance of 0.02237. From Figure D7.5, a 95 percent confidence interval for σ^2 would be

$$(0.016 , 0.037) \quad .$$

Taking square roots gives us the following 95 percent confidence interval for σ:

$$(0.13 , 0.19) \quad .$$

FIGURE D7.4   Output from SEST in butterfat content example

$\langle n \rangle$ SEST

Which of the following parameters do you want to estimate :
1. The mean
2. The proportion
3. The variance or standard deviation
4. Differences between two means
5. Differences between two proportions
6. Ratios of two variances. (answer 1, 2, 3, 4, 5, or 6) ? 3
Sample size ? 20
Enter the sample variance : .02237
Do you want an error bound/confidence interval for :
1. .90
2. .95
3. .99
Answer 1, 2, or 3 : 2
$0.12897234E-01 <$ population variance $< 0.47569327E-01$

FIGURE D7.5   Output from SEST with $s^2 = 0.02237$ and $n = 50$

⟨n⟩ SEST

Which of the following parameters do you want to estimate :
1. The mean
2. The proportion
3. The variance or standard deviation
4. Differences between two means
5. Differences between two proportions
6. Ratios of two variances. (answer 1, 2, 3, 4, 5, or 6) ? 3
Sample size ? 50
Enter the sample variance : 0.02237
Do you want an error bound/confidence interval for :
1. .90
2. .95
3. .99
Answer 1, 2, or 3 : 2
$0.15974442E-01 <$ population variance $< 0.36919296E-01$

Thus, the larger sample size ($n = 50$ instead of $n = 20$) with the same s^2 would provide a 95 percent confidence interval for σ with width

$$0.19 - 0.13 = 0.06 \text{ percent} \quad,$$

as compared with

$$0.22 - 0.11 = 0.11 \text{ percent}$$

when $n = 20$.

EXERCISES

D7.1.

In Exercise D6.3, find a point estimate and a 95 percent confidence interval for the variance of pulse rate in the population of individuals in the health club's aerobics program.

D7.2.

Use DCHI to find the value of χ^2 that cuts off a probability of 0.05 in the right tail of the distribution when we have 10 degrees of freedom. What value cuts off a probability of 0.05 in the *left* tail of this distribution?

D7.3.

Using SHYP with the data from Table A1.5, test the hypothesis that $\sigma^2 = 5$ against the alternative that $\sigma^2 > 5$ at the $\alpha = 0.05$ level.

D8

Goodness of Fit

Sometimes we are interested not just in a summary measure such as a mean, variance, or proportion, but also in whether a particular model or claim or hypothesis (such as a claim that a population is normally distributed or a claim that all classes are equally likely) provides a "good fit" to a set of sample data. Thus, we use the terminology "goodness of fit test." From a sample, we can determine observed frequencies, which tell us how often each event or value or class actually occurs in the sample. From the hypothesized model, we can determine expected frequencies, which tell us how often we *expect* each event or value or class to occur in the sample. If our hypothesis states that some event or value or class i has probability p_i, then e_i, the expected frequency of that event or value or class in a sample of size n, is

$$e_i = np_i \quad .\tag{D8-1}$$

The question in a goodness of fit test is whether the observed and expected frequencies differ enough to cause us to reject the null hypothesis of a good fit. A measure of how much the observed and expected frequencies differ is the chi-square statistic

$$\chi^2 = \sum_{i=1}^{k} \frac{(o_i - e_i)^2}{e_i} \quad ,\tag{D8-2}$$

where

$$o_i = \text{the observed frequency in class } i,$$
$$e_i = \text{the expected frequency in class } i,$$
$$\text{and } k = \text{the number of classes.}$$

The sampling distribuion of the χ^2-statistic is the chi-square distribution with $k-1$ degrees of freedom. (Remember that k is the number of classes, not the number of observations.) The critical value of χ^2 for the goodness of fit test is found in the right-hand tail of the chi-square distribution. This is a one-tailed test to the right because small values of χ^2 indicate a good fit whereas large values indicate a poor fit.

ISP COMMANDS

The command PDIST can be used to conduct a goodness of fit test. The hypothesized distribution can be binomial, gamma, normal, Poisson, uniform, or any distribution consisting of probabilities that you choose and enter yourself. A histogram of the data and the hypothesized distribution are shown, and the step-by-step calculation of the χ^2-statistic is given. The appropriate number of degrees of freedom is noted, and the critical χ^2-value for $\alpha = 0.05$ is provided. If a different level of α is desired, the command DCHI can be used to find the critical χ^2-value.

SOLVED EXAMPLE

A sample of $n = 140$ family incomes is given in Table D8.1. An economist is working on an econometric model and would like to be able to represent the distribution of family incomes in terms of some theoretical distribution such as the normal distribution. The use of normal distributions and other theoretical distributions is often convenient when building a complex model to represent some aspect of the economy.

Since the normal distribution is convenient to use, suppose that we try to see if the normal curve provides a good fit to the income data. The output from PDIST for this goodness of fit test is shown in Figure D8.1. The computed test statistic is $\chi^2 = 55.752$, which is much greater than the critical χ^2-value of 15.507. Note that the computer uses 11 classes. The mean and standard deviation of the normal distribution are estimated from the data. Thus, we have to subtract 2 degrees of freedom because of the estimation of these two parameters, leaving us with $11 - 1 - 2 = 8$ degrees of freedom.

Since the computed test statistic is much larger than the critical value for the goodness of fit test, it is clear that the normal distribution does *not* provide a good fit for the income data. A glance at the histogram indicates why the fit is not good. The histogram of incomes is skewed to the right, with some large incomes providing a long right-hand tail for the distribution. The normal distribution, in contrast, is symmetric and is not able to represent the skewness that is present in the data.

A theoretical distribution capable of representing some skewness to the right is the gamma distribution. This distribution is an option available in PDIST, and some output from PDIST for a test of the goodness of fit of the gamma distribution to the income data is given in Figure D8.2. This time 12 classes are used and 2 degrees of freedom are subtracted because of the estimation of the mean and standard deviation, giving a total of $12 - 1 - 2 = 9$ degrees of freedom.

It appears from both the chi-square goodness of fit test and the histogram with a gamma distribution superimposed that the gamma distribution provides a better fit to the

TABLE D8.1   A sample of 140 family incomes

37,235	12,048	32,187	33,531	19,322
20,784	61,391	63,281	25,444	26,478
64,891	6,691	31,668	16,787	22,488
27,730	17,906	112,922	20,973	42,505
52,617	56,238	19,748	27,066	14,743
12,188	29,785	28,741	14,085	20,247
19,150	23,378	15,721	16,180	25,473
19,245	23,347	32,438	39,662	49,475
26,455	14,393	30,112	55,346	57,707
29,562	40,739	41,956	12,594	15,211
30,515	55,134	22,351	57,174	6,434
38,583	33,465	19,857	23,647	34,926
24,806	51,670	19,861	38,760	
32,732	19,645	20,643	21,590	
32,830	21,192	29,530	16,931	
19,703	29,407	34,513	57,273	
26,591	23,630	14,524	34,114	
21,752	24,743	40,255	39,818	
15,151	32,478	26,536	22,254	
31,803	44,070	30,084	26,112	
20,513	17,725	12,917	36,649	
27,651	30,895	26,609	20,579	
34,814	37,058	37,547	17,095	
26,868	58,116	36,975	19,595	
11,852	78,131	28,796	52,848	
64,966	23,330	29,979	23,373	
21,163	28,782	12,294	25,048	
20,314	15,590	15,813	23,990	
13,643	21,423	36,915	17,160	
17,496	29,707	125,177	41,115	
16,190	53,417	24,692	28,041	
33,317	18,730	37,780	13,541	

income data than does the normal distribution. However, the computed test statistic of $\chi^2 = 28.123$ is still considerably greater than the critical value of 16.919. Thus, the gamma distribution is not able to represent adequately the distribution of incomes.

Sometimes we find it useful to apply transformations to data in order to arrive at a distribution that is close to a theoretical distribution such as the normal curve. With skewed distributions such as incomes, taking logarithms may remove the skewness and provide a histogram that resembles a normal curve. Using the command TLOG, we can transform the income data by taking the natural logarithm of each of the 140 family incomes in Table D8.1. Some output from PHIST for a goodness of fit test of the normal distribution to these log incomes is presented in Figure D8.3. The histogram of log incomes appears relatively symmetric, and the computed χ^2-statistic, 8.297, is much less

FIGURE D8.1   Output from PDIST for goodness of fit of normal distribution to incomes

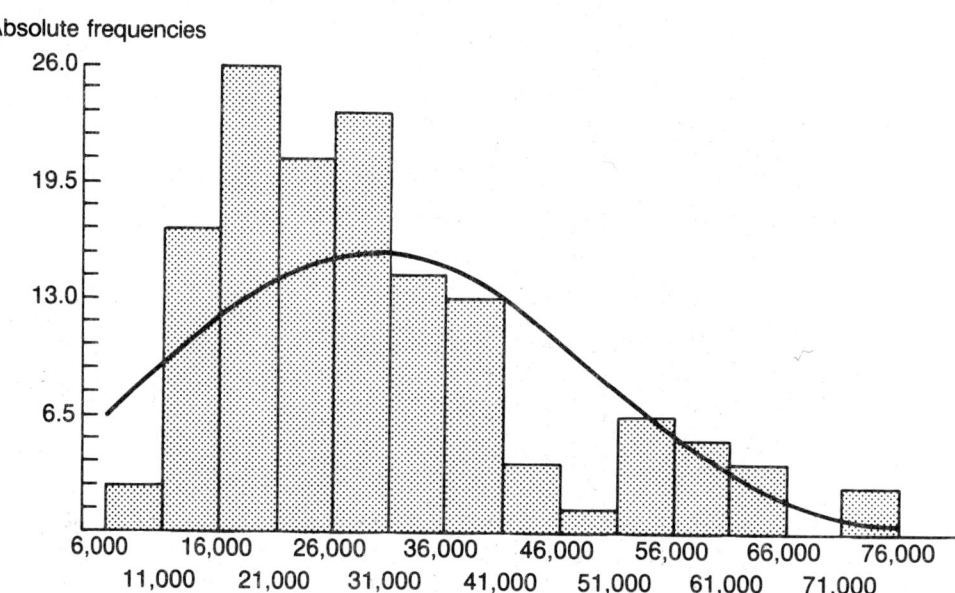

Absolute frequencies

Do you want to see the calculations for the chi-square test ? Y

***** Calculations for Chi-square goodness of fit test *****

Observed (0) Frequencies	Expected (E) Frequencies	$(0-E)$	$(0-E)^2$	$(0-E)^2/E$
2.00	19.68	-17.68	312.6	15.883
17.00	9.95	7.05	49.7	5.000
26.00	12.67	13.33	177.7	14.031
21.00	14.84	6.16	37.9	2.553
24.00	16.00	8.00	63.9	3.995
14.00	15.88	-1.88	3.5	0.222
13.00	14.49	-1.49	2.2	0.154
4.00	12.17	-8.17	66.8	5.489
1.00	9.41	-8.41	70.7	7.516
6.00	6.69	-0.69	0.5	0.072
12.00	9.22	2.78	7.7	0.837

Computed chi-square value = 55.752
Degrees of freedom = 8
Critical chi-square value from table (alpha = 0.05) = 15.507

FIGURE D8.2   Output from PDIST for goodness of fit of gamma distribution to incomes

Absolute frequencies

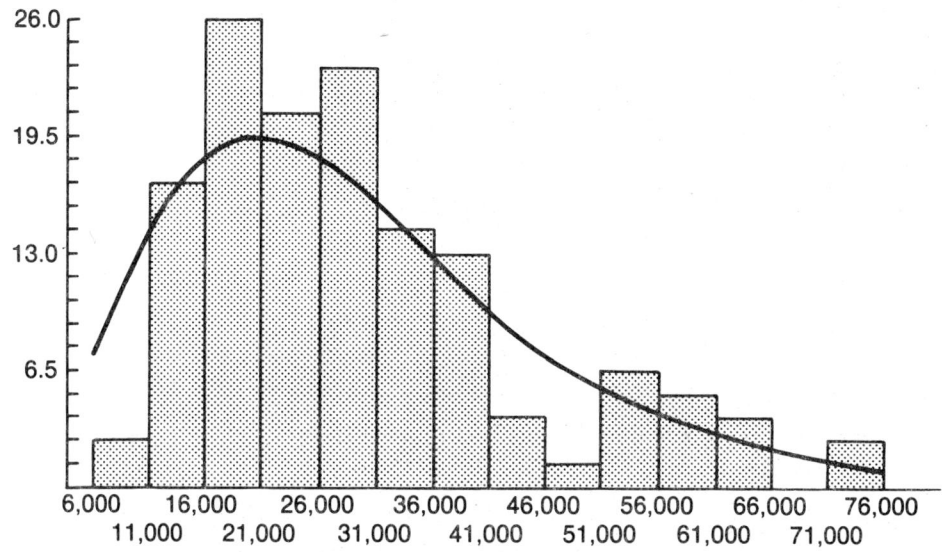

Do you want to see the calculations for the chi-square test ? Y

***** Calculations for chi-square goodness of fit test *****

Observed (O) Frequencies	Expected (E) Frequencies	$(O-E)$	$(O-E)^2$	$(O-E)^2/E$
2.00	16.58	-14.58	212.6	12.821
17.00	16.97	0.03	0.0	0.000
26.00	19.18	6.82	46.5	2.426
21.00	18.62	2.38	5.7	0.304
24.00	16.48	7.52	56.6	3.434
14.00	13.70	0.30	0.1	0.007
13.00	10.89	2.11	4.5	0.411
4.00	8.36	-4.36	19.0	2.275
1.00	6.25	-5.25	27.6	4.414
6.00	4.58	1.42	2.0	0.441
5.00	3.29	1.71	2.9	0.884
7.00	5.10	1.90	3.6	0.706

Computed chi-square value = 28.123
Degrees of freedom = 9
Critical chi-square value from table (alpha = 0.05) = 16.919

FIGURE D8.3   Output from PDIST for goodness of fit of normal distribution to log incomes

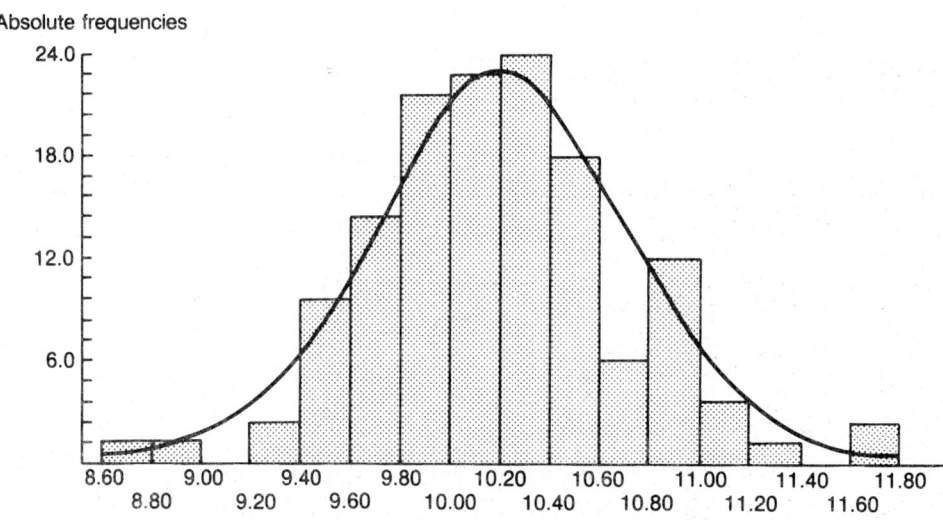

Absolute frequencies

Do you want to see the calculations for the chi-square test ? Y

***** Calculations for chi-square goodness of fit test *****

Observed (0) Frequencies	Expected (E) Frequencies	$(0-E)$	$(0-E)^2$	$(0-E)^2/E$
4.00	6.93	−2.93	8.6	1.238
10.00	8.20	1.80	3.2	0.396
14.00	13.51	0.49	0.2	0.018
22.00	18.86	3.14	9.8	0.522
23.00	22.30	0.70	0.5	0.022
24.00	22.32	1.68	2.8	0.127
18.00	18.92	−0.92	0.8	0.045
6.00	13.58	−7.58	57.4	4.230
12.00	8.25	3.75	14.0	1.700
7.00	7.00	0.00	0.0	0.000

Computed chi-square value = 8.297
Degrees of freedom = 7
Critical chi-square value from table (alpha = 0.05) = 14.067

than the critical value of 14.067. Thus, we can conclude that the normal distribution provides a good fit to the log incomes. The economist might find it convenient to use a normal distribution for log incomes in the econometric model.

EXERCISES

D8.1.

Use PDIST to test, at the $\alpha = 0.05$ level, whether the salaries in Table A1.7 appear to have come from a normal population.

D8.2.

A random number generator in a computer is supposed to generate the 10 digits 0, 1, 2, ..., 9 with equal frequency. In a sample of numbers from the random number generator, 0 appeared 12 times, 1 appeared 8 times, 2 appeared 4 times, 3 appeared 11 times, 4 appeared 15 times, 5 appeared 7 times, 6 appeared 10 times, 7 appeared 12 times, 8 appeared 9 times, and 9 appeared 10 times. Use PDIST to conduct a goodness of fit test of the hypothesis that the digits appear with equal frequency.

Summary of
Part D

In inferential statistics, we use information from a sample to make inferences about the population from which the sample was drawn. Typically such inferences are made in the form of estimates and tests of hypotheses. In Part D, we have described methods for estimation and hypothesis testing involving means, proportions, variances, and even entire distributions (goodness of fit tests). Estimates and tests for comparisons of two or more populations (for example, the comparison of the means of two populations) will be discussed in Part E.

Many different formulas have been presented for specific situations. These various formulas are based on the same general principles of estimation and hypothesis testing, and they draw on concepts and results from Parts A, B, and C. Sample statistics used to describe data sets in Part A have appeared as point estimates in Part D. As emphasized in Part C, the concept of a sampling distribution introduced there is a crucial element in statistical inference. Sampling distributions tell us how large the errors of estimation are likely to be and provide test statistics. The normal distribution from Part B has been encountered frequently, and some new probability distributions have been introduced: the t, chi-square, and beta distributions. In the Bayesian approach to inference, we utilize prior information in addition to sample information, combining the two types of information by using Bayes' theorem, which was discussed in Part B. In general, an appreciation of some basic concepts presented earlier in the book makes it easier to understand what is going on in estimation and hypothesis testing.

VOCABULARY LIST

acceptance region

accepting H_0

alpha

alternative hypothesis

Bayesian estimation

Bayesian interval estimate

Bayesian point estimate

beta

beta distribution

bound on error of estimation

chi-square distribution

chi-square statistic

confidence interval

confidence level

critical value

decision rule

degrees of freedom

error bound

error of estimation

estimate

estimation

estimation of the mean

estimation of the proportion

estimation of the variance

expected frequency

goodness of fit

hypothesis

hypothesis testing

hypothesis testing for the mean

hypothesis testing for the proportion

hypothesis testing for the variance

inferences from large samples

inferences from small samples

inferential statistics

interval estimate

null hypothesis

observed frequency

one-tailed test (one-sided test) to the left

one-tailed test (one-sided test) to the right

point estimate

population proportion

posterior distribution

posterior information

prior distribution

prior information

proportion

rejecting H_0

rejection region

sample mean

sample proportion

sample variance

sampling-theory estimation (classical estimation)

standard error of the mean

standard error of the sample proportion

statistical inference

t distribution (Student t distribution)

t-statistic

test statistic

two-tailed test (two-sided test)

Type I error

Type II error

unbiased estimate

z-statistic

LIST OF SYMBOLS

n	— sample size
μ	— the mean of the population
$\bar{x}$	— the sample mean
s^2	— the sample variance
$\mu_{\bar{x}}$	— the mean of the sampling distribution of $\bar{x}$
$\sigma_{\bar{x}}$	— the standard error of the mean

E	–	a bound on the error of estimation
z	–	a statistic with a standard normal distribution
$z_\alpha(z_{\alpha/2}$, etc.)	–	the value of z with probability α ($\alpha/2$, etc.) to the right in a standard normal distribution
p	–	the population proportion
x/n	–	the sample proportion
$\mu_{x/n}$	–	the mean of the sampling distribution of x/n
$\sigma_{x/n}$	–	the standard error of x/n
p^*	–	a prior guess about p
n'	–	the equivalent sample size for the prior information
m'	–	the mean of the prior distribution
σ'	–	the standard deviation of the prior distribution
n''	–	the equivalent sample size for the posterior distribution
m''	–	the mean of the posterior distribution
σ''	–	the standard deviation of the posterior distribution
x'	–	the equivalent number of ''successes'' for prior information about p
x''	–	the equivalent number of ''successes'' for posterior information about p
H_0	–	a null hypothesis
H_A	–	an alternative hypothesis
α	–	the probability of a Type I error
β	–	the probability of a Type II error
μ_0	–	the value of μ under the null hypothesis
p_0	–	the value of p under the null hypothesis
t	–	a statistic with a t distribution
t_α ($t_{\alpha/2}$, etc.)	–	the value of t with probability α ($\alpha/2$, etc.) to the right in a t distribution
χ^2	–	a statistic with a chi-square distribution
χ^2_α ($\chi^2_{\alpha/2}$, etc.)	–	the value of χ^2 with probability α ($\alpha/2$, etc.) to the right in a chi-square distribution
σ^2_0	–	the value of σ^2 under the null hypothesis
p_i	–	the probability of event i or class i
e_i	–	the expected frequency of event i or class i
o_i	–	the observed frequency of event i or class i
k	–	the number of events or classes in a goodness of fit test

FORMULAS

$$\sigma_{\bar{x}} = \frac{\sigma}{\sqrt{n}}$$

$$s_{\bar{x}} = \frac{s}{\sqrt{n}}$$

$$E = z_{\alpha/2}\left(\frac{\sigma}{\sqrt{n}}\right) \quad \text{or} \quad E = z_{\alpha/2}\left(\frac{s}{\sqrt{n}}\right) \qquad \text{(for estimation of } \mu, \text{ large } n)$$

$$\bar{x} \pm z_{\alpha/2}\left(\frac{\sigma}{\sqrt{n}}\right) \quad \text{or} \quad \bar{x} \pm z_{\alpha/2}\left(\frac{s}{\sqrt{n}}\right) \qquad \text{(confidence interval for } \mu, \text{ large } n)$$

$$\bar{x} \pm E \qquad \text{(confidence interval for } \mu)$$

$$n = \left(\frac{z_{\alpha/2}\,\sigma}{E}\right)^2 \qquad \text{(for estimation of } \mu, \text{ large } n)$$

$$\mu_{x/n} = p$$

$$\sigma_{x/n} = \sqrt{p(1-p)/n}$$

$$E = z_{\alpha/2}\sqrt{\left(\frac{x}{n}\right)\left(1-\frac{x}{n}\right)\Big/n} \qquad \text{(for estimation of } p, \text{ large } n)$$

$$\frac{x}{n} \pm E = \frac{x}{n} \pm z_{\alpha/2}\sqrt{\left(\frac{x}{n}\right)\left(1-\frac{x}{n}\right)\Big/n} \qquad \text{(confidence interval for } p, \text{ large } n)$$

$$n = \frac{z_{\alpha/2}^2}{4E^2} \quad \text{or} \quad n = \frac{z_{\alpha/2}^2\,p^*(1-p^*)}{E^2} \qquad \text{(for estimation of } p, \text{ large } n)$$

$$n' = \left(\frac{\sigma}{\sigma'}\right)^2$$

$$m'' = \frac{n'm' + nm}{n' + n}$$

$$\sigma'' = \frac{\sigma}{\sqrt{n' + n}}$$

$$E = z_{\alpha/2} \left(\frac{\sigma}{\sqrt{n' + n}} \right)$$ (for Bayesian estimation of μ)

$$m'' \pm z_{\alpha/2} \left(\frac{\sigma}{\sqrt{n' + n}} \right)$$ (Bayesian interval estimate for μ)

$$f(p) \frac{(n' - 1)!}{(x' - 1)! \, (n' - x' - 1)!} p^{x' - 1}(1 - p)^{n' - x' - 1}$$ (beta distribution)

$$m' = x'/n'$$ (mean of beta prior distribution)

$$\sigma' = \sqrt{m'(1 - m')/(n' + 1)}$$ (standard deviation of beta prior distribution)

$$n' = \frac{m'(1 - m')}{(\sigma')^2} - 1$$

$$x' = m'n'$$

$$x'' = x' + x$$

$$n'' = n' + n$$

$$m'' = x''/n''$$ (mean of beta posterior distribution)

$$\sigma'' = \sqrt{m''(1 - m'')/(n'' + 1)}$$ (standard deviation of beta posterior distribution)

$$E = z_{\alpha/2} \sqrt{m''(1 - m'')/(n'' + 1)}$$ (for Bayesian estimation of p, large n)

$$\frac{x''}{n''} \pm z_{\alpha/2} \sqrt{m''(1 - m'')/(n'' + 1)}$$ (Bayesian interval estimate for p, large n)

$$z = \frac{\bar{x} - \mu_0}{\sigma/\sqrt{n}} \quad \text{or} \quad z = \frac{\bar{x} - \mu_0}{s/\sqrt{n}}$$ (for tests concerning μ, large n)

$$z = \frac{(x/n) - p_0}{\sqrt{p_0(1 - p_0)/n}}$$ (for tests concerning p, large n)

$$E \quad = t_{\alpha/2} \left(\frac{s}{\sqrt{n}} \right) \qquad\qquad \text{(for estimation of } \mu, \text{ small } n)$$

$$\bar{x} \pm t_{\alpha/2} \left(\frac{s}{\sqrt{n}} \right) \qquad\qquad \text{(confidence interval for } \mu, \text{ small } n)$$

$$t = \frac{\bar{x} - \mu_0}{s/\sqrt{n}} \qquad\qquad \text{(for tests concerning } \mu, \text{ small } n)$$

$$\left(\frac{(n-1)s^2}{\chi^2_{\alpha/2}} \ , \ \frac{(n-1)s^2}{\chi^2_{1-(\alpha/2)}} \right) \qquad\qquad \text{(confidence interval for } \sigma^2)$$

$$\chi^2 \quad = \frac{(n-1)s^2}{\sigma_0^2} \qquad\qquad \text{(for tests concerning } \sigma^2)$$

$$e_i \quad = np_i$$

$$\chi^2 \quad = \sum_{i=1}^{k} \frac{(o_i - e_i)^2}{e_i} \qquad\qquad \text{(for goodness of fit test)}$$

RELEVANT ISP COMMANDS

The most useful commands for Part D are the S commands of ISP. Commands such as SMEAN, SVAR, and STERR can be used to compute point estimates and standard errors. SEST provides error bounds and confidence intervals for means, proportions, variances, and some comparisons that will be discussed in Part E. For these same cases, hypotheses can be tested via SHYP, which provides computed test statistics, critical values, and decisions concerning the acceptance or rejection of hypotheses. The command SBAYE gives posterior distributions and Bayesian point and interval estimates for means and proportions.

The normal, t, and chi-square distributions have been encountered at various spots in Part D. Hence, the following commands from the D group of ISP commands can be helpful by providing probabilities or critical values needed for estimation and hypothesis testing.

DCHI: Chi-square distribution
DNORM: Normal distribution
DTDS: Student t distribution

From the P group of commands, PDIST is particularly valuable for Part D because it can be used to conduct a goodness of fit test. Of course, other P commands are useful, as in previous parts, for printing and plotting, as are C and T commands for arithmetic computations.

The topics covered in Part D are explained briefly in the ME commands. These commands correspond to the eight chapters in Part D.

MED1:	Estimation of the mean for large samples
MED2:	Estimation of the proportion for large samples
MED3:	Bayesian estimation
MED4:	Hypothesis testing for the mean for large samples
MED5:	Hypothesis testing for the proportion for large samples
MED6:	Inferences about the mean for small samples
MED7:	Inferences about the variance
MED8:	Goodness of fit

PART E

Comparing Two or More Populations

In Part D, we discussed inferences for a single population. The situations considered involved a single mean, a single proportion, a single variance, or an entire distribution for a single population. But many statistical studies involve comparisons among two or more populations. In fact, it is probably fair to say that comparisons of populations are more widespread than studies of single populations. We read about comparisons of salaries for different professions, gas mileage for different cars, test scores of students from different schools, effectiveness of different treatments for a disease, and so on.

As in Part D, we will discuss both estimation and hypothesis testing. Furthermore, we will focus on some of the same parameters. However, instead of considering a single mean, for instance, we will look at the difference between two means and make inferences about this difference. The basic notions developed in Part D still apply here, although specific formulas must be developed for each situation.

Chapters E1 and E2 discuss the comparison of two means for large and small samples, respectively. The comparison of two means involves the difference between the means. Two proportions are compared (again in terms of a difference) in Chapter E3. The comparison of two variances, discussed in Chapter E4, involves the ratio of the variances. In Chapters E5 and E6, we describe a method for the comparison of several means. This method is called the analysis of variance. The concluding chapter in Part E, Chapter E7, involves the comparison of several proportions and other situations falling under the study of what statisticians call contingency tables.

E1
Comparing Two Means for Large Samples

In comparing the means of two populations, we usually focus on the difference between the means. The population mean and standard deviation are denoted by μ_1 and σ_1 for the first population and μ_2 and σ_2 for the second population. A random sample of size n_1 is taken from the first population, resulting in a sample mean of $\bar{x}_1$ and sample standard deviation of s_1. A random sample of size n_2 is taken from the second population, with sample mean $\bar{x}_2$ and sample standard deviation s_2. We assume at this point that the samples from the two populations are independent of each other.

A sensible estimate for the difference in means, $\mu_1 - \mu_2$, is simply the corresponding difference in sample means. For large samples the sampling distribution of $\bar{x}_1 - \bar{x}_2$ is approximately normal, and a bound on the error of estimation with probability $1 - \alpha$ is

$$E = z_{\alpha/2} \sqrt{\frac{\sigma_1^2}{n_1} + \frac{\sigma_2^2}{n_2}} \ . \tag{E1--1}$$

A confidence interval for $\mu_1 - \mu_2$ with confidence level $1 - \alpha$ is

$$\bar{x}_1 - \bar{x}_2 \pm z_{\alpha/2} \sqrt{\frac{\sigma_1^2}{n_1} + \frac{\sigma_2^2}{n_2}} \ . \tag{E1--2}$$

For tests concerning $\mu_1 - \mu_2$, the test statistic is

$$z = \frac{\bar{x}_1 - \bar{x}_2 - \delta_0}{\sqrt{\frac{\sigma_1^2}{n_1} + \frac{\sigma_2^2}{n_2}}} \ , \tag{E1--3}$$

where δ_0 represents the value of $\mu_1 - \mu_2$ specified in the null hypothesis H_0.

Sometimes samples from two populations are non-independent in a special way: they are *paired* samples. That is, each observation from one sample is paired in some way with a single observation from the other sample. When samples are paired we can find the individual difference d_i for each pair, and therefore obtain a single sample of n differences that can be analyzed by the methods of Chapters D1 and D4 for single samples. The sample mean and standard deviation of differences are

$$\bar{x}_d = \frac{\sum\limits_{i=1}^{n} d_i}{n} \tag{E1-4}$$

and

$$s_d = \sqrt{\frac{\sum\limits_{i=1}^{n} (d_i - \bar{x}_d)^2}{n - 1}} \; . \tag{E1-5}$$

The estimated standard error of $\bar{x}_d$ is

$$s_{\bar{x}_d} = \frac{s_d}{\sqrt{n}} \tag{E1-6}$$

and a 95 percent confidence interval for $\mu_1 - \mu_2$ is

$$\bar{x}_d \pm z_{\alpha/2} \, s_d/\sqrt{n} \; . \tag{E1-7}$$

For hypothesis tests concerning $\mu_1 - \mu_2$ with paired samples, the test statistic is

$$z = \frac{\bar{x}_d - \delta_0}{s_d/\sqrt{n}} \; , \tag{E1-8}$$

where δ_0 is the value of $\mu_1 - \mu_2$ specified in the null hypothesis H_0.

ISP COMMANDS

The commands SEST and SHYP can be used for estimation and hypothesis testing concerning differences in means. For independent samples, a specific option is available in SEST and SHYP to deal with $\mu_1 - \mu_2$. The summary measures required are the sample size, sample mean, and sample standard deviation (or population standard deviation if it is known) for each sample. These summary measures can be found by entering the two samples in the ISP data matrix and using SMEAN and SVAR.

For paired samples, the samples should be entered and the command TSUB used to find the individual differences. The mean and standard deviation of these differences can be found via SMEAN and SVAR, and the option for dealing with a single mean (the mean of the differences) can be used in SEST or SHYP in order to make inferences about $\mu_1 - \mu_2$.

SOLVED EXAMPLE

Interest rates on home mortgages have shifted considerably in recent years. Thus, the distribution of rates in effect on existing mortgages in a city is influenced by factors relating to when the mortgages were taken out (the number of new houses built in recent years, the rate of turnover in owners of older houses, and so on). In order to compare the mean interest rates on existing mortgages in two cities, random samples of existing mortgages in the two cities are shown in Table E1.1.

The sample mean and standard deviation for the $n_1 = 60$ mortgage rates sampled from City 1 are

$$\bar{x}_1 = 11.48$$

and

$$s_1 = 3.44 \quad .$$

For the $n_2 = 80$ mortgage rates sampled from City 2, we have

$$\bar{x}_2 = 13.82$$

and

$$s_2 = 3.54 \quad .$$

These summary statistics can be calculated easily by using SMEAN and SVAR.

Our point estimate of $\mu_1 - \mu_2$, the difference in mean mortgage rates for all mortgages in the two cities, is

$$\bar{x}_1 - \bar{x}_2 = 11.48 - 13.82 = -2.34 \quad .$$

On average, then, we estimate that the mortgage rates in City 2 are higher than those in City 1 by 2.34 percentage points. Since the population standard deviations σ_1 and σ_2 are not known but the samples are large, we use s_1 and s_2 in computing the standard error of $\bar{x}_1 - \bar{x}_2$:

$$\sqrt{\frac{(3.44)^2}{60} + \frac{(3.54)^2}{80}} = 0.595 \quad .$$

TABLE E1.1   Samples of interest rates on existing mortgages in two cities

City 1 ($n_1 = 60$)				City 2 ($n_2 = 80$)					
6	11	8	10	13	16	11	12	14	14
12	17	9	6	10	13	13	12	15	16
12	3	9	17	15	10	10	9	18	10
12	15	20	9	15	10	18	13	13	14
13	11	8	13	14	7	8	12	18	19
14	11	18	14	17	15	9	20	16	
9	14	15	8	13	9	13	8	16	
9	12	17	11	16	8	20	15	15	
17	12	12	15	14	15	14	15	19	
15	10	7	10	16	23	12	15	18	
11	12	14	10	13	14	16	15	17	
10	12	6	12	8	11	13	12	9	
14	11	13	14	19	16	9	13	12	
10	10	15	11	9	17	14	17	20	
11	4	6	12	11	13	12	10	23	

Thus, with probability 0.99, a bound on the error of estimation in using $\bar{x}_1 - \bar{x}_2$ to estimate $\mu_1 - \mu_2$ is, from (E1–1),

$$E = 2.58 \sqrt{\frac{(3.44)^2}{60} + \frac{(3.54)^2}{80}} = 1.535 \quad .$$

A 99 percent confidence interval for $\mu_1 - \mu_2$ is

$$-2.34 \pm 1.535 \quad ,$$

or

$$(-3.875, \ -0.805) \quad .$$

With confidence level 0.99, we estimate that the mean mortgage rate in City 2 is between 0.805 and 3.875 percentage points higher than the mean mortgage rate in City 1. All these calculations can be handled by SEST, as illustrated in Figure E1.1.

We might simply want to test whether the mean mortgage rates are the same or different in the two cities. The hypotheses are

$$H_0: \mu_1 - \mu_2 = 0$$

and

$$H_A: \mu_1 - \mu_2 \neq 0 \quad ,$$

FIGURE E1.1 Output from SEST for estimation of difference in mean mortgage rates in two cities

⟨n⟩ SEST

Which of the following parameters do you want to estimate :
1. The mean
2. The proportion
3. The variance or standard deviation
4. Differences between two means
5. Differences between two proportions
6. Ratios of two variances. (answer 1, 2, 3, 4, 5, or 6) ? 4
Enter the following information for the 1st population :
Sample size ? 60
Enter the sample mean : 11.48
Enter the standard deviation : 3.44
Enter the following information for the 2nd population :
Sample size ? 80
Enter the sample mean : 13.82
Enter the standard deviation : 3.54
Do you want an error bound/confidence interval for :
1. .90
2. .95
3. .99
Answer 1, 2, or 3 : 3
Difference between the two population means is -2.340 plus/minus 1.535

$-3.875 <$ difference between the two population means < -0.805

and the value of the test statistic is

$$z = \frac{11.48 - 13.82 - 0}{\sqrt{\dfrac{(3.44)^2}{60} + \dfrac{(3.54)^2}{80}}} = -3.934 \quad,$$

as shown in the output from SHYP in Figure E1.2. For $\alpha = 0.01$, the critical values are $z = -2.58$ and $z = 2.58$. Therefore, the computed z-value is in the rejection region, and we conclude that there is a difference between the mean mortgage rates in the two cities.

FIGURE E1.2   Output from SHYP for two-tailed test of difference in mean mortgage rates in two cities

⟨n⟩ SHYP
    ***** Tests Hypotheses *****

Do you want to test a null hypothesis about :
1. The mean
2. The proportion
3. The variance or standard deviation
4. Differences between two means
5. Differences between two proportions
6. Ratios of two variances. (answer 1, 2, 3, 4, 5, or 6) ? 4
Enter the following information for the 1st population :
Sample size ? 60
Enter the sample mean : 11.48
Enter the standard deviation : 3.44
Enter the following information for the 2nd population :
Sample size ? 80
Enter the sample mean : 13.82
Enter the standard deviation : 3.54
Do you want an alpha (Type I error) of :
1. .10
2. .05
3. .01
Answer 1, 2, or 3 : 3
Do you want :
1. A two-tailed test
2. A one-tailed test to the right
3. A one-tailed test to the left (answer 1, 2, or 3) ? 1
The null hypothesis that the difference between the two population means is zero cannot be accepted from the sample evidence
The computed z-value is -3.934
The critical values are -2.580 and 2.580

EXERCISES

E1.1.
    An experiment is conducted to compare two treatments for a particular disease. Patients are randomly assigned to the two treatments, and the length of time until each patient is cured is recorded. For the 40 patients receiving the new treatment, the average time is 3.6 days, with a standard deviation of 1.1 days. For the 45 patients receiving the old treatment, the average time is 4.1 days, with a standard deviation of 1.3 days. Use SHYP to determine whether the evidence is sufficient to allow us to reject the null hypothesis of equal means and to conclude that the new treatment reduces the mean time until a patient is cured.

E1.2.

Students in a French class are given a standardized test at the beginning of the semester and again at the end of the semester. The 45 students average 12 points higher on the post-semester test than on the pre-semester test, and the standard deviation of the difference in scores is 7.4. Use SEST to find a 95 percent confidence interval for the difference in means from the pre-semester test to the post-semester test.

E1.3.

Use GSAMP to generate a sample of size 60 from a normal distribution with mean 80 and standard deviation 10, and save the sample. Then use GSAMP to generate another sample of size 60 from a normal population, this time with mean 90 and standard deviation 10. With SHYP, run a two-tailed test to see if the means appear to be significantly different at the $\alpha = 0.01$ level.

E2
Comparing Two Means for Small Samples

Inferences about $\mu_1 - \mu_2$ with small samples involve t distributions and t-statistics. When the samples are independent, we assume that the two populations are approximately normally distributed and that the population standard deviations are equal. The two sample standard deviations can be pooled to obtained a pooled estimate s_p of the common standard deviation:

$$s_p = \sqrt{\frac{(n_1 - 1)\, s_1^2 + (n_2 - 1)\, s_2^2}{n_1 + n_2 - 2}} \quad . \tag{E2--1}$$

With confidence level $1 - \alpha$, a bound on the error of estimation in using $\bar{x}_1 - \bar{x}_2$ to estimate $\mu_1 - \mu_2$ is

$$E = t_{\alpha/2}\, s_p \sqrt{\frac{1}{n_1} + \frac{1}{n_2}} \quad , \tag{E2--2}$$

and a confidence interval for $\mu_1 - \mu_2$ is of the form

$$\bar{x}_1 - \bar{x}_2 \pm t_{\alpha/2}\, s_p \sqrt{\frac{1}{n_1} + \frac{1}{n_2}} \quad . \tag{E2--3}$$

The test statistic for small-sample tests concerning $\mu_1 - \mu_2$ is

$$t = \frac{\bar{x}_1 - \bar{x}_2 - \delta_0}{s_p \sqrt{\frac{1}{n_1} + \frac{1}{n_2}}} \quad , \tag{E2--4}$$

where δ_0 is the value of $\mu_1 - \mu_2$ specified in H_0.

When the samples are paired, an error bound with probability $1 - \alpha$ is

$$E = t_{\alpha/2} \frac{s_d}{\sqrt{n}} \quad , \tag{E2-5}$$

and the corresponding confidence interval for $\mu_1 - \mu_2$ is

$$\bar{x}_d \pm t_{\alpha/2} \frac{s_d}{\sqrt{n}} \quad . \tag{E2-6}$$

Here the t distribution has $n - 1$ degrees of freedom, as in Chapter D6, because we are now working with a single sample of n differences. For hypothesis testing, we use the test statistic

$$t = \frac{\bar{x}_d - \delta_0}{s_d/\sqrt{n}} \quad . \tag{E2-7}$$

ISP COMMANDS

The commands SEST and SHYP can be used to compare two means for small samples as well as large samples. For independent samples, the option in SEST and SHYP for inferences about differences in means should be used. For paired samples, the differences should be found (with TSUB) and the option in SEST and SHYP for inferences about a single mean should be used. For levels of α other than those available in SHYP or confidence levels other than those available in SEST, DTDS can provide the appropriate t-values.

SOLVED EXAMPLE

A particular task on an assembly line has been performed by the same method for years. Now a new method has been proposed, and it is anticipated that the average time required to do the task is less for the new method than for the old method. A random sample of 12 workers is selected, and each worker performs the task using the old method. The 12 times (in minutes) are recorded, and this group is called the control group. A separate random sample of 12 workers to perform the task using the new method is selected. This group is called the experimental group. The times required to do the task by the members of the control and experimental groups are presented in Table E2.1.

Summary statistics for the two samples are as follows:

$$
\begin{aligned}
n_1 &= 12 \ , & n_2 &= 12 \ , \\
\bar{x}_1 &= 31.17 \ , & \bar{x}_2 &= 28.50 \ , \\
s_1 &= 5.36 \ , & s_2 &= 5.70 \ .
\end{aligned}
$$

TABLE E2.1   Samples of times (in minutes) required to do a task using two methods (control and experimental)

Control group		Experimental group	
24	25	29	37
29	27	29	32
31	36	21	28
33	38	35	33
33	35	25	20
24	39	32	21

The results of a test of

$$H_0: \mu_1 - \mu_2 = 0$$

versus

$$H_A: \mu_1 - \mu_2 > 0$$

via SHYP are shown in Figure E2.1. With $\alpha = 0.05$ and $12 + 12 - 2 = 22$ degrees of freedom, the rejection region for this one-tailed test to the right is $t \geq 1.717$. The computed t-statistic is 1.182. Thus, although the average time required to do the task in the experiment is 2.67 minutes less for the new method than for the old method, the difference is not large enough (given the sample sizes and standard deviations) to convince us that $\mu_2 < \mu_1$. The computation of the t-statistic can be done with (E2–1) and (E2–4):

$$s_p = \sqrt{\frac{(12 - 1)(5.36)^2 + (12 - 1)(5.70)^2}{12 + 12 - 2}} = 5.533$$

$$\text{and } t = \frac{31.17 - 28.50 - 0}{5.533\sqrt{\frac{1}{12} + \frac{1}{12}}} = 1.182 \quad .$$

An alternative way to design the experiment would be to use the same 12 workers for both methods. This would give us paired samples. For example, suppose that we do this and obtain the data shown in Table E2.2. The times for the two methods are recorded and the differences for the 12 workers are computed. The mean and standard deviation of the 12 differences are

$$\bar{x}_d = 3.75$$

FIGURE E2.1 Output from SHYP for comparison of mean times required to do a task using two methods

⟨n⟩ SHYP

***** Tests Hypotheses *****

Do you want to test a null hypothesis about :
1. The mean
2. The proportion
3. The variance or standard deviation
4. Differences between two means
5. Differences between two proportions
6. Ratios of two variances. (answer 1, 2, 3, 4, 5, or 6) ? 4
Enter the following information for the 1st population :
Sample size ? 12
Enter the sample mean : 31.17
Enter the standard deviation : 5.36
Enter the following information for the 2nd population :
Sample size ? 12
Enter the sample mean : 28.50
Enter the standard deviation : 5.70
Do you want an alpha (Type I error) of :
1. .10
2. .05
3. .01
Answer 1, 2, or 3 : 2
Do you want :
1. A two-tailed test
2. A one-tailed test to the right
3. A one-tailed test to the left (answer 1, 2, or 3) ? 2
The null hypothesis that the difference between the two population means is zero can be accepted
The computed t-value is 1.182
The critical value is 1.717

and

$$s_d = 3.77 \quad .$$

Now we can use SHYP with the option for testing hypotheses about a single mean, $\mu_d = \mu_1 - \mu_2$. The results for a test of

$$H_0: \mu_1 - \mu_2 = 0$$

versus

$$H_A: \mu_1 - \mu_2 > 0$$

TABLE E2.2 Paired samples of times (in minutes) required to do a task using two methods

WORKER	OLDM	NEWM	DIF
* 1*	24.000	23.000	1.000
* 2*	29.000	25.000	4.000
* 3*	31.000	28.000	3.000
* 4*	33.000	29.000	4.000
* 5*	33.000	26.000	7.000
* 6*	24.000	26.000	−2.000
* 7*	25.000	25.000	0.000
* 8*	27.000	28.000	−1.000
* 9*	36.000	29.000	7.000
* 10*	38.000	30.000	8.000
* 11*	35.000	31.000	4.000
* 12*	39.000	29.000	10.000
Sum	374.0000	329.0000	45.0000
n =	12	12	12
Mean	31.1667	27.4167	3.7500

with $\alpha = 0.05$ are shown in Figure E2.2. The test statistic is

$$t = \frac{3.75 - 0}{3.77/\sqrt{12}} = 3.446$$

with $12 - 1 = 11$ degrees of freedom. This value is greater than the critical value, implying that we should reject the hypothesis that $\mu_1 - \mu_2 = 0$ and conclude that the mean time required to do the job is lower for the new method than for the old method.

EXERCISES

E2.1.

The same test is given to students from an honors algebra class and students from a regular algebra class. The scores for the students in the honors class are 82, 76, 59, 95, 83, 98, 93, and 91. The scores for the students in the regular class are 89, 65, 75, 79, 94, 89, 81, 77, 72, 87, and 90. Assuming equal variances in the two classes, test to see whether the mean for honors students is higher than the mean for regular students, as opposed to the null hypothesis in which the two means are equal.

E2.2.

In Exercise E1.2, suppose that there were only 12 students in the French class, and the differences in scores were as follows (a positive difference indicates that the student performed better on the post-semester test): 15, −8, 7, 13, 9, 2, −5, 12, 0, 6, −1,

FIGURE E2.2   Output from SHYP for comparison of mean times required to do a task using two methods (paired samples)

⟨n⟩ SHYP
       ***** Tests Hypotheses *****

Do you want to test a null hypothesis about :
1. The mean
2. The proportion
3. The variance or standard deviation
4. Differences between two means
5. Differences between two proportions
6. Ratios of two variances. (answer 1, 2, 3, 4, 5, or 6) ? 1
Sample size ? 12
Enter the population mean from the null hypotheses : 0
Enter the sample mean : 3.75
Enter the standard deviation : 3.77
Do you want an alpha (Type I error) of :
1. .10
2. .05
3. .01
Answer 1, 2, or 3 : 2
Do you want :
1. A two-tailed test
2. A one-tailed test to the right
3. A one-tailed test to the left (answer 1, 2, or 3) ? 2
The sample evidence suggests that the null hypothesis that the mean is zero should be rejected
The computed t-value is 3.446
The critical value is 1.796

and 9. Use SEST to find a 95 percent confidence interval for the difference in means from the pre-semester test to the post-semester test.

E2.3.

Use GSAMP to generate two samples, each of size 5, from a normal population with mean 50 and standard deviation 10. Use a two-tailed test to see whether the null hypothesis of equality of means can be rejected at the $\alpha = 0.10$ level. Then repeat the entire procedure 19 more times. In the 20 tests, how many times were you able to reject the null hypothesis? Discuss your results.

E3
Comparing Two Proportions

In this chapter we will consider inferences about a difference in proportions, $p_1 - p_2$, from two different populations. We assume that random samples independent of each other are taken from the two populations. The sample size is n_1 for the sample from the first population, and x_1 denotes the number of times the event of interest (for example, an item is defective, a patient is cured) occurs. For the second sample, x_2 represents the number of times the event of interest occurs and n_2 represents the sample size.

A reasonable estimate of the difference in population proportions, $p_1 - p_2$, is simply the corresponding difference in sample proportions,

$$\frac{x_1}{n_1} - \frac{x_2}{n_2} \ .$$

With probability $1 - \alpha$, a bound on the error of estimation when $(x_1/n_1) - (x_2/n_2)$ is used to estimate $p_1 - p_2$ is

$$E = z_{\alpha/2} \sqrt{\frac{\dfrac{x_1}{n_1}\left(1 - \dfrac{x_1}{n_1}\right)}{n_1} + \frac{\dfrac{x_2}{n_2}\left(1 - \dfrac{x_2}{n_2}\right)}{n_2}} \ . \qquad \text{(E3–1)}$$

The corresponding confidence interval with confidence level $1 - \alpha$ is

$$\frac{x_1}{n_1} - \frac{x_2}{n_2} \pm z_{\alpha/2} \sqrt{\frac{\dfrac{x_1}{n_1}\left(1 - \dfrac{x_1}{n_1}\right)}{n_1} + \frac{\dfrac{x_2}{n_2}\left(1 - \dfrac{x_2}{n_2}\right)}{n_2}} \ . \qquad \text{(E3–2)}$$

For a test of

$$H_0 : p_1 - p_2 = 0$$

against a one-tailed or two-tailed alternative, the test statistic is

$$z = \frac{\dfrac{x_1}{n_1} - \dfrac{x_2}{n_2}}{\sqrt{\left(\dfrac{x_1 + x_2}{n_1 + n_2}\right)\left(1 - \dfrac{x_1 + x_2}{n_1 + n_2}\right)\left(\dfrac{1}{n_1} + \dfrac{1}{n_2}\right)}} . \tag{E3-3}$$

ISP COMMANDS

The commands SEST and SHYP offer an option for estimation and hypothesis testing involving differences between two proportions. For confidence levels and values of α not available in SEST and SHYP, the appropriate z-values can be found by using DNORM.

SOLVED EXAMPLE

In recent years diet drinks have become popular as people have become concerned about controlling their weight. Much of the advertising for diet drinks has been aimed at females, reflecting a feeling that females are more likely than males to purchase diet drinks. An experiment designed to investigate possible differences between males and females in preferences for soft drinks offers an individual a choice between a regular cola and a diet cola. In a sample of $n_1 = 300$ males, $x_1 = 192$ chose the regular cola and the remaining 108 selected the diet cola. In a sample of $n_2 = 300$ females, $x_2 = 144$ picked the regular cola and the remaining 156 took the diet cola.

The sample proportions choosing the regular cola are

$$x_1/n_1 = 192/300 = 0.64$$

for the males and

$$x_2/n_2 = 144/300 = 0.48$$

for the females. As anticipated, a higher proportion of males selected the regular (non-diet) cola. Our estimate of $p_1 - p_2$, the difference in proportions choosing the regular cola in the entire populations of males and females, is

$$\frac{x_1}{n_1} - \frac{x_2}{n_2} = 0.64 - 0.48 = 0.16 .$$

That is, we estimate that the proportion of males choosing regular cola is 0.16 higher than the proportion of females choosing regular cola.

The estimated standard error of

$$\frac{x_1}{n_1} - \frac{x_2}{n_2}$$

is

$$\sqrt{\frac{0.64\,(0.36)}{300} + \frac{0.48\,(0.52)}{300}} = 0.04 \quad .$$

Thus, as shown in Figure E3.1, a bound with probability 0.99 on the error of estimation in using $(x_1/n_1) - (x_2/n_2)$ to estimate $p_1 - p_2$ is, from (E3–1),

$$E = 2.58 \sqrt{\frac{0.64\,(0.36)}{300} + \frac{0.48\,(0.52)}{300}} = 2.58\,(0.04) = 0.103 \quad .$$

FIGURE E3.1 Output from SEST for comparison of proportions of males and females selecting regular cola over diet cola

⟨n⟩ SEST

Which of the following parameters do you want to estimate :
1. The mean
2. The proportion
3. The variance or standard deviation
4. Differences between two means
5. Differences between two proportions
6. Ratios of two variances. (answer 1, 2, 3, 4, 5, or 6) ? 5
Enter the following information for the 1st population :
Sample size ? 300
Enter the sample proportion : 0.64
Enter the following information for the 2nd population :
Sample size ? 300
Enter the sample proportion : 0.48
Do you want an error bound/confidence interval for :
1. .90
2. .95
3. .99
Answer 1, 2, or 3 : 3
Difference between the two population proportions is 0.160 plus/minus 0.103

0.057 < difference between the two population proportions < 0.263

A 99 percent confidence interval for $p_1 - p_2$ is

$$0.16 \pm 0.103 \quad,$$

or

$$(0.057, 0.263) \quad .$$

We estimate with 99 percent confidence that the proportion of males choosing regular cola over diet cola is between 0.057 and 0.263 higher than the proportion of females choosing regular cola over diet cola.

For a test of

$$H_0 : p_1 - p_2 = 0$$

versus

$$H_A : p_1 - p_2 > 0 \quad,$$

the test statistic in (E3–3) takes on the value

$$z = \frac{0.64 - 0.48}{\sqrt{\left(\dfrac{192 + 144}{300 + 300}\right)\left(1 - \dfrac{192 + 144}{300 + 300}\right)\left(\dfrac{1}{300} + \dfrac{1}{300}\right)}} = 3.948 \quad,$$

as calculated via SHYP in Figure E3.2. If $\alpha = 0.05$, the critical value for a one-tailed test to the right is $z = 1.64$. Thus, we should reject the hypothesis of no difference between proportions. It appears that males and females do differ in their preferences between regular cola and diet cola.

EXERCISES

E3.1.

In a metropolitan area, the inner city voters have traditionally voted Democratic and the suburban voters have voted Republican. A pre-election poll shows that 123 of 200 randomly chosen inner city voters and 137 of 300 randomly chosen suburban voters prefer the Democratic candidate. Find a 99 percent confidence interval for the difference between the proportion of inner city voters and the proportion of suburban voters favoring the Democratic candidate.

FIGURE E3.2   Output from SHYP for comparison of proportions of males and females selecting regular cola over diet cola

⟨n⟩ SHYP

       ***** Tests Hypotheses *****

Do you want to test a null hypothesis about :
1.   The mean
2.   The proportion
3.   The variance or standard deviation
4.   Differences between two means
5.   Differences between two proportions
6.   Ratios of two variances. (answer 1, 2, 3, 4, 5, or 6) ? 5
Enter the following information for the 1st population :
Sample size ? 300
Enter the sample proportion : 0.64
Enter the following information for the 2nd population :
Sample size ? 300
Enter the sample proportion : 0.48
Do you want an alpha (Type I error) of :
1.   .10
2.   .05
3.   .01
Answer 1, 2, or 3 : 2
Do you want :
1.   A two-tailed test
2.   A one-tailed test to the right
3.   A one-tailed test to the left (answer 1, 2, or 3) ? 2
The null hypothesis that the difference between the two population proportions is zero cannot be accepted from the sample evidence
The computed z-value is 3.948
The critical value is 1.640

E3.2.

    In London, a sample of 200 people yields 35 left-handers. In Paris, a sample of 150 people yields 17 left-handers. Is this evidence strong enough to cause us to reject the hypothesis that the proportion of left-handers is the same in London and Paris, using an alpha of 0.05?

E4
Comparing Two Variances

Because of the nature of the sampling distributions involved, it is more convenient when comparing variances to work with a ratio of variances, σ_1^2/σ_2^2, instead of a difference in variances. If the samples from the two populations are independent of each other, and if the two populations are normally distributed, then a distribution known as an F distribution can be used to make inferences about σ_1^2/σ_2^2. In particular, the statistic

$$F = \frac{s_1^2/\sigma_1^2}{s_2^2/\sigma_2^2} \qquad \text{(E4–1)}$$

has an F distribution with $n_1 - 1$ degrees of freedom in the numerator and $n_2 - 1$ degrees of freedom in the denominator. Specific F-values cutting off certain right-tail probabilities can be found in tables of the F distribution.

The general form for a confidence interval for σ_1^2/σ_2^2 with confidence level $1 - \alpha$ is

$$\left(\frac{s_1^2/s_2^2}{F_{\alpha/2}} \; , \; \frac{s_1^2/s_2^2}{F_{1-(\alpha/2)}} \right) \; , \qquad \text{(E4–2)}$$

where $F_{\alpha/2}$ is the value cutting off probability $\alpha/2$ in the right tail of the F distribution with $n_1 - 1$ and $n_2 - 1$ degrees of freedom, and $F_{1-(\alpha/2)}$ cuts off probability $1 - (\alpha/2)$ to the right. Note that the higher F-value is used to compute the lower limit of the confidence interval and the lower F-value is used in the calculation of the upper limit. For a test of

$$H_0: \sigma_1^2 = \sigma_2^2$$

against the alternative

$$H_A: \sigma_1^2 \neq \sigma_2^2 \; ,$$

the test statistic is

$$F = \frac{s_1^2}{s_2^2} \quad . \tag{E4-3}$$

ISP COMMANDS

The commands SEST and SHYP can be used for estimation and hypothesis testing concerning ratios of variances. The sample sizes and sample standard deviations (which can be calculated via SVAR) are needed. To get specific F-values or tables of F-values, use DFDS.

SOLVED EXAMPLE

Various types of mutual funds are available for investors. Suppose that we would like to compare the variability in return from two different types of funds. The first type consists of mutual funds investing in blue-chip common stocks, and the second type consists of funds investing in corporate bonds. We want to test whether the variability of return is the same for both types of funds or whether there is more variability among stock funds than among bond funds.

Random samples of $n_1 = 25$ blue-chip stock funds and $n_2 = 16$ corporate bond funds are taken, and the return for each fund (expressed in terms of an annual percentage return) is found for a specified period of time. The returns are given in Table E4.1. The sample means are

$$\bar{x}_1 = 8.22 \text{ and } \bar{x}_2 = 7.20 \quad ,$$

TABLE E4.1 Annual rates of return for $n_1 = 25$ stock funds and $n_2 = 16$ bond funds

Stock funds			Bond funds	
7.4	7.6	8.2	8.4	7.8
10.8	9.4	8.6	6.5	7.1
4.7	5.5	9.0	8.8	4.2
6.3	11.0	6.5	9.3	6.3
13.1	4.1	7.7	7.1	8.0
9.8	9.6	9.8	4.6	9.1
9.2	10.7	8.1	7.5	5.2
10.4	9.5	9.9	8.4	6.9
−1.3				

and the sample standard deviations are

$$s_1 = 2.88 \text{ and } s_2 = 1.54 \quad .$$

As expected, the stock funds have a higher mean return but also have a higher standard deviation.

The hypotheses of interest are

$$H_0: \sigma_1^2 = \sigma_2^2$$

and

$$H_A: \sigma_1^2 > \sigma_2^2 \quad .$$

If $\alpha = 0.05$, the critical value of F (with $25 - 1 = 24$ and $16 - 1 = 15$ degrees of freedom) is 2.29 from Table E4.2. The computed test statistic, using (E4–3), is

$$F = \frac{(2.88)^2}{(1.54)^2} = 3.50 \quad ,$$

as given in the output from SHYP in Figure E4.1. Since this is greater than the critical value for this one-tailed test to the right, we conclude that the variance among the blue-chip common stock funds is greater than the variance among the corporate bond funds.

A confidence interval computed from (E4–2) can give us an idea about how extreme the ratio σ_1^2/σ_2^2 is likely to be. From Table E4.2 with 24 and 15 degrees of freedom,

$$F_{0.05} = 2.29 \quad .$$

With 15 and 24 degrees of freedom, $F_{0.05} = 2.11$. This implies that with 24 and 15 degrees of freedom,

$$F_{0.95} = \frac{1}{2.11} = 0.474 \quad .$$

A 90 percent confidence interval for σ_1^2/σ_2^2 is therefore

$$\left(\frac{(2.88)^2/(1.54)^2}{2.29} \quad , \quad \frac{(2.88)^2/(1.54)^2}{0.474} \right) \quad ,$$

or (1.53, 7.38), as shown in the output from SEST in Figure E4.2. In terms of standard deviations, we have

TABLE E4.2 A table of F-values for a right-tail probability of 0.05

D.f. for denom.	Degrees of freedom for numerator																		
	1	2	3	4	5	6	7	8	9	10	12	15	20	24	30	40	60	120	∞
1	161.4	199.5	215.7	224.6	230.2	234.0	236.8	238.9	240.5	241.9	243.9	245.9	248.0	249.1	250.1	251.1	252.2	253.3	243.3
2	18.51	19.00	19.16	19.25	19.30	19.33	19.35	19.37	19.38	19.40	19.41	19.43	19.45	19.45	19.46	19.47	19.48	19.49	19.50
3	10.13	9.55	9.28	9.12	9.01	8.94	8.89	8.85	8.81	8.79	8.74	8.70	8.66	8.64	8.62	8.59	8.57	8.55	8.53
4	7.71	6.94	6.59	6.39	6.26	6.16	6.09	6.04	6.00	5.96	5.91	5.86	5.80	5.77	5.75	5.72	5.69	5.66	5.63
5	6.61	5.79	5.41	5.19	5.05	4.95	4.88	4.82	4.77	4.74	4.68	4.62	4.56	4.53	4.50	4.46	4.43	4.40	4.36
6	5.99	5.14	4.76	4.53	4.39	4.28	4.21	4.15	4.10	4.06	4.00	3.94	3.87	3.84	3.81	3.77	3.74	3.70	3.67
7	5.59	4.74	4.35	4.12	3.97	3.87	3.79	3.73	3.68	3.64	3.57	3.51	3.44	3.41	3.38	3.34	3.30	3.27	3.23
8	5.32	4.46	4.07	3.84	3.69	3.58	3.50	3.44	3.39	3.35	3.28	3.22	3.15	3.12	3.08	3.04	3.01	2.97	2.93
9	5.12	4.26	3.86	3.63	3.48	3.37	3.29	3.23	3.18	3.14	3.07	3.01	2.94	2.90	2.86	2.83	2.79	2.75	2.71
10	4.96	4.10	3.71	3.48	3.33	3.22	3.14	3.07	3.02	2.98	2.91	2.85	2.77	2.74	2.70	2.66	2.62	2.58	2.54
11	4.84	3.98	3.59	3.36	3.20	3.09	3.01	2.95	2.90	2.85	2.79	2.72	2.65	2.61	2.57	2.53	2.49	2.45	2.40
12	4.75	3.89	3.49	3.26	3.11	3.00	2.91	2.85	2.80	2.75	2.69	2.62	2.54	2.51	2.47	2.43	2.38	2.34	2.30
13	4.67	3.81	3.41	3.18	3.03	2.92	2.83	2.77	2.71	2.67	2.60	2.53	2.46	2.42	2.38	2.34	2.30	2.25	2.21
14	4.60	3.74	3.34	3.11	2.96	2.85	2.76	2.70	2.65	2.60	2.53	2.46	2.39	2.35	2.31	2.27	2.22	2.18	2.13
15	4.54	3.68	3.29	3.06	2.90	2.79	2.71	2.64	2.59	2.54	2.48	2.40	2.33	2.29	2.25	2.20	2.16	2.11	2.07
16	4.49	3.63	3.24	3.01	2.85	2.74	2.66	2.59	2.54	2.49	2.42	2.35	2.28	2.24	2.19	2.15	2.11	2.06	2.01
17	4.45	3.59	3.20	2.96	2.81	2.70	2.61	2.55	2.49	2.45	2.38	2.31	2.23	2.19	2.15	2.10	2.06	2.01	1.96
18	4.41	3.55	3.16	2.93	2.77	2.66	2.58	2.51	2.46	2.41	2.34	2.27	2.19	2.15	2.11	2.06	2.02	1.97	1.92
19	4.38	3.52	3.13	2.90	2.74	2.63	2.54	2.48	2.42	2.38	2.31	2.23	2.16	2.11	2.07	2.03	1.98	1.93	1.88
20	4.35	3.49	3.10	2.87	2.71	2.60	2.51	2.45	2.39	2.35	2.28	2.20	2.12	2.08	2.04	1.99	1.95	1.90	1.84
21	4.32	3.47	3.07	2.84	2.68	2.57	2.49	2.42	2.37	2.32	2.25	2.18	2.10	2.05	2.01	1.96	1.92	1.87	1.81
22	4.30	3.44	3.05	2.82	2.66	2.55	2.46	2.40	2.34	2.30	2.23	2.15	2.07	2.03	1.98	1.94	1.89	1.84	1.78
23	4.28	3.42	3.03	2.80	2.64	2.53	2.44	2.37	2.32	2.27	2.20	2.13	2.05	2.01	1.96	1.91	1.86	1.81	1.76
24	4.26	3.40	3.01	2.78	2.62	2.51	2.42	2.36	2.30	2.25	2.18	2.11	2.03	1.98	1.94	1.89	1.84	1.79	1.73
25	4.24	3.39	2.99	2.76	2.60	2.49	2.40	2.34	2.28	2.24	2.16	2.09	2.01	1.96	1.92	1.87	1.82	1.77	1.71
26	4.23	3.37	2.98	2.74	2.59	2.47	2.39	2.32	2.27	2.22	2.15	2.07	1.99	1.95	1.90	1.85	1.80	1.75	1.69
27	4.21	3.35	2.96	2.73	2.57	2.46	2.37	2.31	2.25	2.20	2.13	2.06	1.97	1.93	1.88	1.84	1.79	1.73	1.67
28	4.20	3.34	2.95	2.71	2.56	2.45	2.36	2.29	2.24	2.19	2.12	2.04	1.96	1.91	1.87	1.82	1.77	1.71	1.65
29	4.18	3.33	2.93	2.70	2.55	2.43	2.35	2.28	2.22	2.18	2.10	2.03	1.94	1.90	1.85	1.81	1.75	1.70	1.64
30	4.17	3.32	2.92	2.69	2.53	2.42	2.33	2.27	2.21	2.16	2.09	2.01	1.93	1.89	1.84	1.79	1.74	1.68	1.62
40	4.08	3.23	2.84	2.61	2.45	2.34	2.25	2.18	2.12	2.08	2.00	1.92	1.84	1.79	1.74	1.69	1.64	1.58	1.51
60	4.00	3.15	2.76	2.53	2.37	2.25	2.17	2.10	2.04	1.99	1.92	1.84	1.75	1.70	1.65	1.59	1.53	1.47	1.39
120	3.92	3.07	2.68	2.45	2.29	2.17	2.09	2.02	1.96	1.91	1.83	1.75	1.66	1.61	1.55	1.50	1.43	1.35	1.25
∞	3.84	3.00	2.60	2.37	2.21	2.10	2.01	1.94	1.88	1.83	1.75	1.67	1.57	1.52	1.46	1.39	1.32	1.22	1.00

FIGURE E4.1   Some output from SHYP for the mutual-fund example

⟨n⟩ SHYP

***** Tests Hypotheses *****

Do you want to test a null hypothesis about :
1.   The mean
2.   The proportion
3.   The variance or standard deviation
4.   Differences between two means
5.   Differences between two proportions
6.   Ratios of two variances. (answer 1, 2, 3, 4, 5, or 6) ? 6
Enter the following information for the 1st population :
Sample size ? 25
Enter the standard deviation : 2.88
Enter the following information for the 2nd population :
Sample size ? 16
Enter the standard deviation : 1.54
Do you want :
1.   A two-tailed test
2.   A one-tailed test to the right
3.   A one-tailed test to the left (answer 1, 2, or 3) ? 2
Do you want an alpha (Type I error) of :
1.   .05
2.   .01
Answer 1 or 2 : 1
The null hypothesis that the ratio of the two population variances is equal cannot be accepted
The critical F-value (24, 15 d.f.) is 2.29
The computed F-value is   3.497

$$(\sqrt{1.53},\ \sqrt{7.38}) = (1.24,\ 2.72)\quad .$$

We are 90 percent confident that the population standard deviation of return for the stock funds is between 1.24 and 2.72 times as large as the population standard deviation of return for the bond funds.

FIGURE E4.2 Some output from SEST for the mutual-fund example

⟨n⟩ SEST

***** Computes Point and Interval Estimates *****

Which of the following parameters do you want to estimate :
1. The mean
2. The proportion
3. The variance or standard deviation
4. Differences between two means
5. Differences between two proportions
6. Ratios of two variances. (answer 1, 2, 3, 4, 5, or 6) ? 6
Enter the following information for the 1st population :
Sample size ? 25
Enter the standard deviation : 2.88
Enter the following information for the 2nd population :
Sample size ? 16
Enter the standard deviation : 1.54
Do you want an error bound/confidence interval for :
1. .90
2. .98
Answer 1 or 2 : 1

1.53   <Ratio of two population variances <   7.38

EXERCISES

E4.1.
    Use SHYP to test (with $\alpha = 0.05$) whether the variances are the same in the two classes in Exercise E2.1.

E4.2.
    In Exercise E1.1, use SEST to find a 90 percent confidence interval for the ratio of the variance for the new treatment to the variance for the old treatment.

E4.3.
    From FTDS, find the F-value that cuts off a probability of 0.05 in the right-hand tail of an F distribution with 8 and 15 degrees of freedom.

E5
Comparing Several Means

Comparisons need not be restricted to differences between the means of two populations. In this chapter and the next chapter, we will discuss inferences that involve several means. We assume that we are dealing with random samples of size n from each of J populations and that the samples are independent of each other. Since we are interested in comparing population means, we might start by calculating the J sample means. If the J samples are combined, the average of all nJ observations is called the grand sample mean $\bar{x}$. By subtracting this grand sample mean from each of the J individual sample means (denoted by $\bar{x}_1, \bar{x}_2, \ldots, \bar{x}_J$), we find the estimated treatment effects $\bar{x}_1 - \bar{x}, \bar{x}_2 - \bar{x}, \ldots, \bar{x}_J - \bar{x}$. Are any differences in treatment effects due only to sampling fluctuations, or are they also due in part to differences among the J population means $\mu_1, \mu_2, \ldots, \mu_J$? The general idea underlying the statistical comparison of several means involves looking at variability. We know that the variability *within* each sample is attributable to sampling fluctuations. The variability *between* samples, however, can be attributable in part to any differences among the population means as well as to sampling fluctuations. Thus, if the variability between samples appears to be significantly greater than the variability within samples, this can be taken as evidence that the population means are not equal. The formal statistical test that uses this notion of comparing variability involves an F-statistic that will be presented in the next chapter.

ISP COMMANDS

Individual sample means and standard deviations can be found by using SMEAN and SVAR, and standard errors can be calculated with STERR. The grand sample mean and the standard deviation of the J sample means can be determined by using SMEAN and SVAR on the J sample means.

SOLVED EXAMPLE

In the example in Chapter E2 involving the amount of time required to do a task, average times were compared for two methods. We will generalize this example and compare the mean times for four methods instead of just two. Four separate random samples of $n = 12$ workers are selected, and each of the four groups is assigned one of the four methods. The first group uses the old method and the second group uses the new method considered in the example in Chapter E2. Thus, the data in Table E5.1 for the first two groups are identical to the data in Table E2.1. Groups 3 and 4 use two other new methods and provide the rest of the data in Table E5.1.

We can see from the means at the bottom of Table E5.1 that the average times differ for the four groups:

$$\bar{x}_1 = 31.17 \quad ,$$

$$\bar{x}_2 = 28.50 \quad ,$$

$$\bar{x}_3 = 23.33 \quad ,$$

$$\text{and } \bar{x}_4 = 32.25 \quad .$$

The grand sample mean is

$$\bar{x} = \frac{31.17 + 28.50 + 23.33 + 32.25}{4} = 28.81 \quad .$$

The estimated treatment effects are

$$\bar{x}_1 - \bar{x} = 31.17 - 28.81 = 2.36 \quad ,$$
$$\bar{x}_2 - \bar{x} = 28.50 - 28.81 = -0.31 \quad ,$$
$$\bar{x}_3 - \bar{x} = 23.33 - 28.81 = -5.48 \quad ,$$
$$\text{and } \bar{x}_4 - \bar{x} = 32.25 - 28.81 = 3.44 \quad .$$

The average time for the third method is 5.48 minutes below the grand mean, and the worst method appears to be the fourth, with an average time 3.44 minutes above the grand mean.

The variances and standard deviations for the four groups are given in Figure E5.1. The standard deviations are

$$s_1 = 5.36 \quad ,$$
$$s_2 = 5.70 \quad ,$$
$$s_3 = 4.64 \quad ,$$
$$\text{and } s_4 = 7.66 \quad .$$

TABLE E5.1   Samples of times (in minutes) required to do a task using four methods

	GRP1	GRP2	GRP3	GRP4
	24.000	29.000	22.000	39.000
	29.000	29.000	22.000	28.000
	31.000	21.000	23.000	40.000
	33.000	35.000	33.000	27.000
	33.000	25.000	29.000	30.000
	24.000	32.000	19.000	20.000
	25.000	37.000	24.000	36.000
	27.000	32.000	17.000	22.000
	36.000	28.000	21.000	42.000
	38.000	33.000	18.000	34.000
	35.000	20.000	26.000	42.000
	39.000	21.000	26.000	27.000
Sum	374.0000	342.0000	280.0000	387.0000
n	12	12	12	12
Mean	31.1667	28.5000	23.3333	32.2500

FIGURE E5.1   Output from SVAR for times from four methods

***** Computes the Variance and Standard Deviation *****

Which column (s) do you want to work on ? 1, 2, 3, 4
Here are the results for your 12 numbers:

				Variance	Standard deviation
Column	1	(GRP1	) :	28.696970	5.3569555
Column	2	(GRP2	) :	32.454544	5.6968889
Column	3	(GRP3	) :	21.515152	4.6384430
Column	4	(GRP4	) :	58.750000	7.6648550

These are the sample variance and standard deviation—i.e., the divisor is $n - 1$.

The corresponding standard errors of the mean are

$$s_1/\sqrt{n_1} = 5.36/\sqrt{12} = 1.55 \quad,$$
$$s_2/\sqrt{n_2} = 5.70/\sqrt{12} = 1.65 \quad,$$
$$s_3/\sqrt{n_3} = 4.64/\sqrt{12} = 1.34 \quad,$$
$$\text{and } s_4/\sqrt{n_4} = 7.66/\sqrt{12} = 2.21 \quad.$$

In contrast, the standard deviation of the four sample means is

$$\sqrt{\frac{(31.17-28.81)^2 + (28.50-28.81)^2 + (25.33-28.81)^2 + (32.25-28.81)^2}{4-1}} \quad,$$

which equals 3.98. The individual standard errors of the mean range from 1.34 to 2.21. The fact that the standard deviation of the four sample means is much higher suggests that the variation among the sample means might be due not just to sampling fluctuations but also to differences among the four population means μ_1, μ_2, μ_3, and μ_4. We will continue this example and present a test of differences among the four population means in the next chapter.

EXERCISES

E5.1.

An experiment concerning the output per hour for four machines gives the results shown in Table E5.2. Find the sample means and sample variances for the four machines, estimate the treatment effects, and compare the standard deviation of the four sample means with the individual standard errors for the four machines. Discuss the results.

TABLE E5.2   Hourly output for four machines

Machine 1	Machine 2	Machine 3	Machine 4
160	134	104	86
155	139	175	71
170	144	96	112
175	150	83	110
152	156	89	87
167	159	79	100
180	170	84	105
154	133	86	93
141	128	83	85

E5.2.

Use GSAMP to generate 3 samples of size 6 from a normal population with mean 50 and standard deviation 10. Estimate the treatment effects and compare the standard deviation of the three sample means with the three individual standard errors.

E5.3.

Repeat Exercise E5.2, but let the standard deviations be 5, 10, and 15 for the 3 samples instead of all being 10. Compare the results with those found in Exercise E5.2.

E6
Analysis of Variance

The analysis of variance is a technique for comparing the variability within samples and the variability between samples. As in the previous chapter, we assume that we are dealing with random samples of size n from J populations and that the samples are independent of each other. If the ith observation in the jth sample is denoted by x_{ij}, then the sum of squared deviations from the grand sample mean, called the total sum of squares, is

$$\text{SS total} = \sum_{j=1}^{J} \sum_{i=1}^{n} (x_{ij} - \bar{x})^2 \quad . \tag{E6-1}$$

The double summation sign indicates that we are summing over all n observations within each sample *and* over all J samples (hence, over all nJ observations). Since we are dealing with nJ observations, the number of degrees of freedom associated with SS total is $nJ - 1$. The total sum of squares can be separated into two components, a sum of squares between groups and a sum of squares within groups. These are

$$\text{SS between} = n \sum_{j=1}^{J} (\bar{x}_j - \bar{x})^2 \quad , \tag{E6-2}$$

with $J - 1$ degrees of freedom, and

$$\text{SS within} = \sum_{j=1}^{J} \left[\sum_{i=1}^{n} (x_{ij} - \bar{x}_j)^2 \right] \quad , \tag{E6-3}$$

with $J(n - 1)$ degrees of freedom. Sometimes SS within is referred to as SS error. It is always true that SS total = SS between + SS within.

Estimates of variability between groups and within groups are

$$\text{MS between} = \frac{\text{SS between}}{J - 1} \tag{E6-4}$$

TABLE E6.1 Analysis of variance table

Source of variation	d.f.	Sum of squares	Mean square
Between groups	$J - 1$	SS between	MS between
Within groups (error)	$J(n - 1)$	SS within	MS within
Total	$nJ - 1$	SS total	

and

$$\text{MS within} = \frac{\text{SS within}}{J(n - 1)} \, . \qquad \text{(E6–5)}$$

The degrees of freedom, sums of squares, and mean squares can be summarized conveniently in an analysis of variance table, as shown in Table E6.1. The test of the hypothesis of equality of means,

$$H_0: \mu_1 = \mu_2 = \mu_3 = \cdots = \mu_J \;\; ,$$

utilizes the statistic

$$F = \frac{\text{MS between}}{\text{MS within}} \;\; , \qquad \text{(E6–6)}$$

which has $J - 1$ degrees of freedom in the numerator and $J(n - 1)$ degrees of freedom in the denominator. If the means are unequal, that will tend to lead to greater variability between groups than would be expected with equal means. This higher variability would translate into a higher value of MS between and therefore to a higher value of the F-statistic. Therefore, the rejection region is located in the right tail of the F distribution.

ISP COMMANDS

The ISP command RANOV performs the computations required for an analysis of variance, provides an analysis of variance table, computes the F-statistic, and gives the critical F-value at the $\alpha = 0.05$ level. The command DFDS can be used to generate a table of F-values or a particular F-value.

SOLVED EXAMPLE

We will now continue the analysis of the data involving the times required to do a task using four different methods. The data are given in Table E5.1, with $J = 4$ groups and $n = 12$ times per group. The analysis of variance table generated by RANOV is displayed

FIGURE E6.1   Output from RANOV for comparison of mean times from four methods of doing a task

⟨n⟩ RANOV

***** Analysis of Variance *****

Need help ? N
Which column(s) do you want to work on ? 1, 2, 3, 4

	d.f.	Sum of Squares	Mean Square
Between groups	3	569.7290	189.90967
Within groups (error)	44	1555.5835	35.35417
Total	47	2125.3125	

Computed F   = 5.3716340
d.f. for numerator   = 3
d.f. for denominator = 44
Value of F from table (alpha = 0.05) = 2.82

in Figure E6.1. The computed F-statistic is 5.37, which is considerably higher than the critical F-value for $\alpha = 0.05$ with 3 and 44 degrees of freedom. Thus, we can reject the hypothesis

$$H_0: \mu_1 = \mu_2 = \mu_3 = \mu_4$$

and conclude that the mean times required to do the task are not the same for the four methods. The third method, which has the lowest average time in the experiment (see Table E5.1), looks the most promising.

EXERCISES

E6.1.
    Use RANOV to perform an analysis of variance on the data in Table E5.2 concerning hourly output for four machines.

E6.2.
    In Exercises E5.2 and E5.3, use RANOV to perform an analysis of variance. Compare the results of the two analyses and discuss.

E7

Contingency Tables

When we take a sample in order to study a proportion, the sample tells us how many times the event of interest occurred and how many times it did not occur. These are observed frequencies of occurrence and non-occurrence. In the comparison of several proportions, we have several such sets of observed frequencies, which can be displayed in a two-dimensional array called a contingency table. As in the case of goodness of fit tests, we would like to compare the observed frequencies with expected frequencies. Under the hypothesis that the proportions are all equal, the expected frequency for a cell in a contingency table can be found as follows:

$$\text{Expected frequency} = \frac{(\text{row total})\,(\text{column total})}{\text{grand total}} \quad . \qquad \text{(E7--1)}$$

The test statistic for a test of equality of proportions is a χ^2-statistic like that used in goodness of fit tests (Chapter D8):

$$\chi^2 = \Sigma\, \frac{(o_i - e_i)^2}{e_i} \quad , \qquad \text{(E7--2)}$$

where o_i and e_i represent observed and expected frequencies in cell i of the contingency table and the summation is taken over all cells of the table. The sampling distribution of the χ^2-statistic in (E7--2) is a chi-square distribution with $(r - 1)(c - 1)$ degrees of freedom, where r represents the number of rows in the contingency table and c represents the number of columns. As is the case with the chi-square goodness of fit test, the rejection region is in the upper tail of the chi-square distribution.

Contingency tables are not just used to deal with simple proportions. If, instead of a simple dichotomy (an event occurs or it does not occur), we have several possible events or classes, then the contingency table will have more than two rows, with (E7--1) and (E7--2) still being applicable. Another useful generalization is that the same approach can be used if the contingency table represents a single sample classified in two dimensions instead of a series of separate samples. In this case the test is sometimes thought of as a test of independence of the two dimensions.

ISP COMMANDS

The expected frequencies can be found by using the command CALC in ISP. Then, if the observed frequencies and the corresponding expected frequencies are entered as two separate variables, the χ^2-statistic can be calculated by using TSUB (subtracting the expected frequencies from the observed frequencies), TCC (squaring the differences), and TDIV (dividing the squared differences by the expected frequencies). Critical χ^2-values can be found by using DCHI.

SOLVED EXAMPLE

In an example in Chapter E3, we compared the proportions of males and females choosing regular cola over diet cola. Here we will discuss a related experiment. There are now three choices: regular cola, diet cola, and caffeine-free cola. Also, instead of comparing males and females, we are comparing three age groups: under 20 years old, 20-39 years old, and 40 and older. Samples of 100 in the youngest age group and 150 in each of the other age groups are taken. Each person is asked to choose a single drink from the three available choices. The resulting observed frequencies are shown in Table E7.1.

Using (E7–1), we can compute the expected frequencies shown in Table E7.2. For example, in the upper left-hand corner of the contingency table, the expected frequency of individuals under 20 years of age choosing regular cola is

$$\frac{152\,(100)}{400} = 38 \quad .$$

To calculate the chi-square statistic given by (E7–2), we can use ISP commands. We enter the 9 observed frequencies and 9 expected frequencies as separate variables,

TABLE E7.1 Contingency table with observed frequencies in age-drink preference example

	Age			
	Under 20	20–40	40 and older	Totals
Regular cola	49	45	58	152
Diet cola	28	46	38	112
Caffeine-free cola	23	59	54	136
Totals	100	150	150	400

TABLE E7.2   Contingency table with expected frequencies in age-drink preference example

	Age			
	Under 20	20–40	40 and older	Totals
Regular cola	38	57	57	152
Diet cola	28	42	42	112
Caffeine-free cola	34	51	51	136
Totals	100	150	150	400

TABLE E7.3   Summary of calculation of chi-square statistic in age-drink preference example

Obsfr	Expfr	Dif	Difsq	Chisq
49.000	38.000	11.000	121.000	3.184
45.000	57.000	− 12.000	144.000	2.526
58.000	57.000	1.000	1.000	0.018
28.000	28.000	0.000	0.000	0.000
46.000	42.000	4.000	16.000	0.381
38.000	42.000	−4.000	16.000	0.381
23.000	34.000	− 11.000	121.000	3.559
59.000	51.000	8.000	64.000	1.255
54.000	51.000	3.000	9.000	0.176
400.0000	400.0000	0.0000	492.0000	11.4802

use TSUB to find the nine $(o_i - e_i)$ differences, use TCC to square these differences, and use TDIV to divide each difference by the corresponding e_i. The results of these steps are shown in Table E7.3. The chi-square statistic is the sum of the last column, 11.48.

Is $\chi^2 = 11.48$ large enough to reject the hypothesis that the three age groups do not differ in their preferences among regular cola, diet cola, and caffeine-free cola? We have

$$(r - 1)(c-1) = (3 - 1)(3 - 1) = 2(2) = 4$$

degrees of freedom. For $\alpha = 0.05$, the critical value of χ^2 is 9.488, and for $\alpha = 0.01$, the critical value is 13.277. Thus, the null hypothesis would be rejected at the $\alpha = 0.05$ level but not at the $\alpha = 0.01$ level.

EXERCISES

E7.1.

A firm sells materials to other firms and is concerned with overdue accounts. Some data regarding overdue accounts from small, medium, and large firms are given in Table E7.4. Is the evidence sufficient for us to conclude that the proportion of accounts that are overdue varies among the three firm sizes?

Table E7.4   Overdue accounts from small, medium, and large firms

	Small firms	Medium firms	Large firms
Accounts overdue	26	15	10
Accounts not overdue	38	34	38

E7.2.

Three candidates are running in an election. A pre-election poll is taken in the three districts voting in this election. In District 1, 45 voters prefer Candidate A, 52 prefer Candidate B, and 31 prefer Candidate C. In District 2, 28 prefer A, 41 prefer B, and 25 prefer C. In District 3, 47 prefer A, 40 prefer B, and 31 prefer C. Test to see whether the three districts differ significantly (at the $\alpha = 0.05$ level) in their voting preferences among the three candidates.

Summary of Part E

Part E represents a continuation of the discussion of inferential statistics presented in Part D. Part D is concerned with inferences about a single population, and Part E involves inferences comparing two or more populations. The same basic concepts of estimation and hypothesis testing are encountered in both Parts D and E. Thus, although the formulas are different (sometimes a bit messier) when two or more populations are involved, the underlying ideas are the same. Furthermore, as noted at the beginning of Part E, it is probably fair to say that comparisons of populations are more widespread than studies of single populations. Many important real-world questions boil down to comparisons: "Is one better than the other, and if so, by how much?"

Moving from statistical analysis for one sample from one population to comparisons involving two or more samples from two or more populations greatly expands the set of real-world problems with which we can deal. However, considering just one variable at a time (gasoline mileage, starting salary, IQ, rate of return on an investment) still restricts us somewhat. For many real-world problems, several variables are of interest. At the end of Part E, we have introduced the notion of relationships among variables by presenting a statistical test for independence in contingency tables. This brief introduction sets the stage for Part F, which involves the statistical analysis of relationships through methods known as correlation and regression. The analysis of relationships among variables further expands our toolbox of statistical techniques.

VOCABULARY LIST

analysis of variance

analysis of variance table

cell in a contingency table

comparing populations

comparing several means

comparing several proportions

comparing two means

comparing two proportions

comparing two variances

contingency table

degrees of freedom for denominator

degrees of freedom for numerator

difference in means

difference in proportions

estimation of a difference in means

estimation of a difference in proportions

estimation of a ratio of variances

F distribution

F statistic

grand mean

hypothesis testing for a difference in means

hypothesis testing for a difference in proportions

hypothesis testing for a ratio of variances

hypothesis testing for contingency tables

hypothesis testing in analysis of variance

independent samples

mean square between groups

mean square within groups

paired samples

pooled estimate of a proportion

pooled estimate of a standard deviation

ratio of variances

standard error of a difference in means

standard error of a difference in proportions

sum of squares between groups

sum of squares within groups

test of independence in contingency tables

total sum of squares

treatment effects

variability between samples

variability within samples

LIST OF SYMBOLS

n_i	— the sample size for the ith sample
$\bar{x}_i$	— the sample mean for the ith sample
s_i	— the sample standard deviation for the ith sample
μ_i	— the mean of the ith population
σ_i	— the standard deviation of the ith population
$\bar{x}_1 - \bar{x}_2$	— a difference in sample means
$\mu_1 - \mu_2$	— a difference in population means
$\mu_{\bar{x}_1 - \bar{x}_2}$	— the mean of the sampling distribution of $\bar{x}_1 - \bar{x}_2$
$\sigma_{\bar{x}_1 - \bar{x}_2}$	— the standard error of $\bar{x}_1 - \bar{x}_2$
δ_0	— the value of $\mu_1 - \mu_2$ under the null hypothesis
d_i	— the ith difference in a paired sample
$\bar{x}_d$	— the sample mean of the differences
μ_d	— the population mean difference
s_d	— the sample standard deviation of the differences
p_i	— the population proportion for the ith population
x_i	— the number of times the event of interest occurs in the ith sample
x_i/n_i	— the sample proportion for the ith sample
$(x_1/n_1) - (x_2/n_2)$	— a difference in sample proportions
s_1^2/s_2^2	— a ratio of sample variances

σ_1^2/σ_2^2	–	a ratio of population variances
J	–	the number of populations or samples
$\bar{x}$	–	the grand sample mean in analysis of variance
$x_i - \bar{x}$	–	the estimated treatment effect for the ith population
F_α ($F_{\alpha/2}$, etc.)	–	the value of F with probability α ($\alpha/2$, etc.) to the right in an F distribution
o_i	–	the observed frequency in cell i of a contingency table
e_i	–	the expected frequency in cell i of a contingency table
r	–	the number of rows in a contingency table
c	–	the number of columns in a contingency table

FORMULAS

$$\mu_{\bar{x}_1 - \bar{x}_2} = \mu_1 - \mu_2$$

$$\sigma_{\bar{x}_1 - \bar{x}_2} = \sqrt{\frac{\sigma_1^2}{n_1} + \frac{\sigma_2^2}{n_2}}$$

$$e = z_{\alpha/2} \sqrt{\frac{\sigma_1^2}{n_1} + \frac{\sigma_2^2}{n_2}} \quad \text{or} \quad E = z_{\alpha/2} \sqrt{\frac{s_1^2}{n_1} + \frac{s_2^2}{n_2}}$$

(for estimation of $\mu_1 - \mu_2$, independent samples, large n_1 and n_2)

$$\bar{x}_1 - \bar{x}_2 \pm z_{\alpha/2} \sqrt{\frac{\sigma_1^2}{n_1} + \frac{\sigma_2^2}{n_2}} \quad \text{or} \quad \bar{x}_1 - \bar{x}_2 \pm z_{\alpha/2} \sqrt{\frac{s_1^2}{n_1} + \frac{s_2^2}{n_2}}$$

(confidence interval for $\mu_1 - \mu_2$, independent samples, large n_1 and n_2)

$$z = \frac{\bar{x}_1 - \bar{x}_2 - \delta_0}{\sqrt{\frac{\sigma_1^2}{n_1} + \frac{\sigma_2^2}{n_2}}} \quad \text{or} \quad z = \frac{\bar{x}_1 - \bar{x}_2 - \delta_0}{\sqrt{\frac{s_1^2}{n_1} + \frac{s_2^2}{n_2}}}$$

(for tests concerning $\mu_1 - \mu_2$, independent samples, large n_1 and n_2)

$$\bar{x}_d = \frac{\sum\limits_{i=1}^{n} d_i}{n}$$

$$s_d = \sqrt{\frac{\sum\limits_{i=1}^{n} (d_i - \bar{x}_d)^2}{n - 1}}$$

$$z = \frac{\bar{x}_d - \delta_0}{s_d/\sqrt{n}}$$

(for tests concerning $\mu_1 - \mu_2$, paired samples, large n)

$$s_p = \sqrt{\frac{(n_1 - 1)s_1^2 + (n_2 - 1)s_2^2}{n_1 + n_2 - 2}}$$

$$E = t_{\alpha/2}\, s_p \sqrt{\frac{1}{n_1} + \frac{1}{n_2}}$$

(for estimation of $\mu_1 - \mu_2$, independent samples, small samples)

$$\bar{x}_1 - \bar{x}_2 \pm t_{\alpha/2}\, s_p \sqrt{\frac{1}{n_1} + \frac{1}{n_2}}$$

(confidence interval for $\mu_1 - \mu_2$, independent samples, small samples)

$$t = \frac{\bar{x}_1 - \bar{x}_2 - \delta_0}{s_p \sqrt{\dfrac{1}{n_1} + \dfrac{1}{n_2}}}$$

(for tests concerning $\mu_1 - \mu_2$, independent samples, small samples)

$$E = t_{\alpha/2}\, \frac{s_d}{\sqrt{n}}$$

(for estimation of $\mu_1 - \mu_2$, paired samples, small n)

$$\bar{x}_d \pm t_{\alpha/2}\, \frac{s_d}{\sqrt{n}}$$

(confidence interval for $\mu_1 - \mu_2$, paired samples, small n)

$$t = \frac{\bar{x}_d - \delta_0}{s_d/\sqrt{n}}$$

(for tests concerning $\mu_1 - \mu_2$, paired samples, small n)

$$E = z_{\alpha/2} \sqrt{\frac{\dfrac{x_1}{n_1}\left(1 - \dfrac{x_1}{n_1}\right)}{n_1} + \frac{\dfrac{x_2}{n_2}\left(1 - \dfrac{x_2}{n_2}\right)}{n_2}}$$

(for estimation of $p_1 - p_2$, large samples)

$$\frac{x_1}{n_1} - \frac{x_2}{n_2} \pm z_{\alpha/2} \sqrt{\frac{\dfrac{x_1}{n_1}\left(1 - \dfrac{x_1}{n_1}\right)}{n_1} + \frac{\dfrac{x_2}{n_2}\left(1 - \dfrac{x_2}{n_2}\right)}{n_2}}$$

(confidence interval for $p_1 - p_2$, large samples)

$$z = \frac{\dfrac{x_1}{n_1} - \dfrac{x_2}{n_2}}{\sqrt{\left(\dfrac{x_1 + x_2}{n_1 + n_2}\right)\left(\dfrac{1}{n_1} + \dfrac{1}{n_2}\right)}}$$

(for test of $p_1 = p_2$, large samples)

$$F = \frac{s_1^2/\sigma_1^2}{s_2^2/\sigma_2^2}$$

$$\left(\frac{s_1^2/s_2^2}{F_{\alpha/2}} \quad , \quad \frac{s_1^2/s_2^2}{F_{1-(\alpha/2)}}\right)$$

(confidence interval for σ_1^2/σ_2^2)

$$F = \frac{s_1^2}{s_2^2}$$

(for test of $\sigma_1^2 = \sigma_2^2$)

$$\text{SS total} = \sum_{j=1}^{J} \sum_{i=1}^{n} (x_{ij} - \bar{x})^2$$

$$\text{SS between} = n \sum_{j=1}^{J} (\bar{x}_j - \bar{x})^2$$

$$\text{SS within} = \sum_{j=1}^{J} \left[\sum_{i=1}^{n} (x_{ij} - \bar{x}_j)^2 \right]$$

$$\text{SS total} = \text{SS between} + \text{SS within}$$

$$\text{MS between} = \frac{\text{SS between}}{J - 1}$$

$$\text{MS within} = \frac{\text{SS within}}{J(n - 1)}$$

$$F = \frac{\text{MS between}}{\text{MS within}}$$

(for test of equality of means in analysis of variance)

$$\chi^2 = \Sigma \frac{(o_i - e_i)^2}{e_i}$$

(for tests of equality of proportions or independence in a contingency table)

$$\text{Expected frequency} = \frac{(\text{row total}) (\text{column total})}{\text{grand total}}$$

RELEVANT ISP COMMANDS

As in Part D, SEST and SHYP are especially useful for much of Part E. These commands can be used for estimation and hypothesis testing involving differences in means, differences in proportions, and ratios of variances. For the purposes of analysis of variance, the command RANOV does the necessary calculations and displays an analysis of variance table. Of course, commands such as SMEAN, SVAR, and STERR are helpful for the computation of various statistics that are used in the formulas encountered in Part E. As usual, the C and T commands can also be helpful.

Inferences in Part E involve the normal, t, F, and chi-square distributions. Therefore, the following commands from the D groups of commands can be helpful by providing probabilities or critical values needed for estimation and hypothesis testing.

DCHI: Chi-square distribution
DFDS: *F* distribution
DNORM: Normal distribution
DTDS: Student *t* distribution

The MC commands provide brief explanations of the concepts discussed in Part E. The seven MC commands correspond to the seven chapters in Part E.

MCE1:   Comparing two means for large samples
MCE2:   Comparing two means for small samples
MCE3:   Comparing two proportions
MCE4:   Comparing two variances
MCE5:   Comparing several means
MCE6:   Analysis of variance
MCE7:   Contingency tables

PART F

Regression and Correlation

Many statistical studies involve relationships among variables. We may want to know whether two variables (for instance, number of cigarettes smoked and incidence of lung cancer) are statistically related. Or, we may know the value of one variable and may want to use this information to help us predict a second variable (for instance, using a score on an aptitude test to predict academic performance). Extending this notion, we may want to use information about several variables to help us predict yet another variable.

To deal with situations such as these, we need to be able to say something about the form of the relationship between two (or among several) variables and also about the strength of the relationship. In terms of form, many relationships can be approximated by a linear function, while others are nonlinear. As for strength, we know that some variables are very closely related whereas others are only slightly related. In Part F we will discuss statistical techniques for studying and making inferences about relationships among variables. We will also describe how we can take advantage of these relationships when we want to make predictions.

Chapter F1 provides a general discussion of functional relationships, and Chapter F2 describes two summary measures, the covariance and correlation coefficient, that tell us something about the direction and strength of the relationship between two variables. Expressing one variable as a linear function of another in order to make predictions is called simple linear regression, and this topic is covered in Chapters F3, F4, and F5. The restriction to linear functions is dropped in Chapter F6, and the restriction to two variables is dropped in the discussion of multiple regression in Chapter F7.

F1
Functional Relationships

When variables are independent, then information about one of the variables does not tell us anything about the other variable. At the other extreme from independence, two variables may be related so strongly that knowing the value of one of the variables enables us to figure out the exact value of the second variable. Such relationships are called functions. If knowing x lets us find y, then y is a function of x. In symbols, we write

$$y = f(x) \quad . \qquad\qquad\qquad\qquad (F1\text{--}1)$$

Examples of functions include linear, quadratic, logarithmic, and exponential functions.

In statistics we deal with situations in which there is some variability or uncertainty. Often variables may be related in some manner but do not have a perfect functional relationship. To learn about the relationship between two variables, we gather information by taking a sample of *pairs* of values (such as the height-weight pair for a given person). A scatter diagram, which is a display of the pairs of values on a two-dimensional graph, can be very useful in giving us an impression of the nature and strength of the relationship between two variables. Summary measures that tell us more about the relationship will be presented in the next chapter.

ISP COMMANDS

The command PSCAT will provide a scatter diagram for data on any pair of variables. As for functions, using transformations—such as squaring (TCC), taking a square root (TSQT), taking logarithms (TLOG), or making an exponential transformation (TEXP)—and then plotting a scatter diagram of the original and transformed variables via PSCAT can help to illustrate the shapes of some types of functions.

SOLVED EXAMPLE

An argument often made in favor of staying in school and getting more education is that people with more education tend to get better jobs and enjoy higher incomes. Is there indeed a relationship between amount of education and income? The data in Table F1.1 are based on a random sample of 40-year-olds. For each person in the sample, the number of years of education and the current income are given.

Obviously there is not a perfect relationship between years of education and income. For example, the second and third individuals in the sample have the same amount of education (12 years), but one earns $10,200 more than the other. Also, the last individual in the sample has less education than the next-to-last individual (14 years, as compared to 16) but has a higher income. However, it does seem that there is a tendency for more education to be associated with a higher income.

The scatter diagram in Figure F1.1 helps us better understand any relationship that may exist between education and income. Although there is a considerable amount of variability, it appears that those individuals with more years of education tend to have higher incomes.

TABLE F1.1 Years of education and income for a random sample of 15 40-year-olds

Observ.	Education	Income
* 1*	13.000	28900.000
* 2*	12.000	35400.000
* 3*	12.000	25200.000
* 4*	10.000	32300.000
* 5*	12.000	32900.000
* 6*	15.000	40600.000
* 7*	8.000	25600.000
* 8*	16.000	34000.000
* 9*	10.000	22200.000
* 10*	13.000	27900.000
* 11*	14.000	38600.000
* 12*	11.000	29100.000
* 13*	16.000	38200.000
* 14*	16.000	30700.000
* 15*	14.000	35700.000
Sum	192.0000	477300.0000
n =	15	15
Mean	12.8000	31820.0000

FIGURE F1.1   A scatter diagram of the incomes and years of education for a sample of 14 40-year-olds

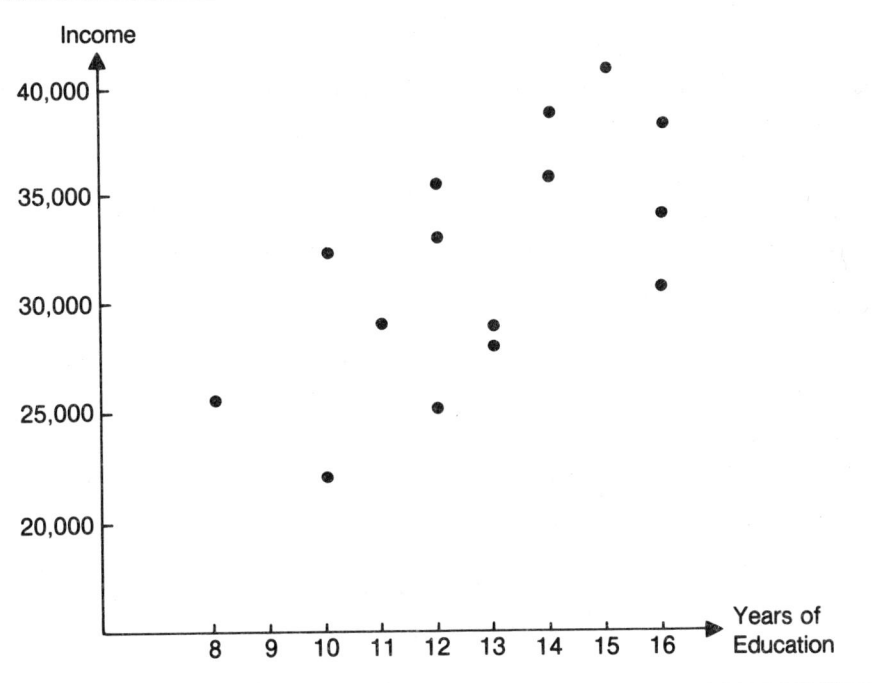

EXERCISES

F1.1.

A statistician is interested in the relationship between the height (in inches) and the weight (in pounds) of students from a particular school. A random sample of students is taken, with the results given in Table F1.2. Use PSCAT to construct a scatter diagram for this set of data.

TABLE F1.2   Heights and weights of 11 students

Height	Weight	Height	Weight	Height	Weight
70	175	75	198	64	156
67	180	71	178	70	182
68	160	76	204	68	167
69	169	70	162		

F1.2.

Enter a data set consisting of the integers from 1 to 20 in ISP. Then use the command TCC to square these values and PSCAT to plot a scatter diagram of the original values paired with their squares. Repeat this process with TSQT, TLOG, and TEXP in place of TCC. Comment on the differences among the four scatter diagrams.

F1.3.

A utility providing electricity is interested in the relationship between the high temperature (in degrees Fahrenheit) and the peak load on the system (in megawatts). A sample of 12 days provides the data in Table F1.3. Use PSCAT to construct a scatter diagram with high temperature on the horizontal axis and peak load on the vertical axis.

TABLE F1.3   High temperature (°F) and peak load (megawatts) for 12 days

Temp.	Peak load	Temp.	Peak load	Temp.	Peak load
92	215	86	189	96	238
88	223	94	223	90	251
82	217	88	194	95	243
91	241	85	223	89	219

F2
Covariance and Correlation

If we take a random sample of n pairs of observations on two variables x and y, then the sample consists of (x_1, y_1), (x_2, y_2), ..., (x_n, y_n). One measure that tells us something about the relationship between x and y is the sample covariance, denoted by $\text{cov}(x,y)$:

$$\text{cov}(x,y) = \frac{\sum_{i=1}^{n} (x_i - \bar{x})(y_i - \bar{y})}{n - 1} \quad . \tag{F2-1}$$

The sign of the covariance indicates whether the relationship is positive (as one variable increases, the other also tends to increase) or negative (as one increases, the other tends to decrease). A measure that also provides information about the strength of the relationship is the sample correlation coefficient, denoted by r:

$$r = \frac{\text{cov}(x,y)}{s_x \, s_y} \quad . \tag{F2-2}$$

The correlation coefficient is a standardized measure, with

$$-1 \leqslant r \leqslant 1 \quad . \tag{F2-3}$$

The magnitude of r indicates the strength of the linear relationship between the variables, with $r = 1$ for a perfect positive linear relationship and $r = -1$ for a perfect negative linear relationship. The closer r is to zero, the weaker the linear relationship. When two variables are independent, their correlation is zero. For computational purposes, an alternative formula for r is

$$r = \frac{n\Sigma x_i y_i - (\Sigma x_i)(\Sigma y_i)}{\sqrt{[n\Sigma x_i^2 - (\Sigma x_i)^2][n\Sigma y_i^2 - (\Sigma y_i)^2]}} \quad , \tag{F2-4}$$

where all summations are from $i = 1$ to $i = n$.

The correlation coefficient r is based on sample data. It is an estimate of the population correlation coefficient, which is denoted by ρ (Greek rho). To test whether x and y are at all linearly related, we test

$$H_0: \rho = 0$$

versus

$$H_A: \rho \neq 0 \quad .$$

The test statistic, based on the assumption that x and y are jointly normally distributed, is

$$t = \frac{r\sqrt{n-2}}{\sqrt{1-r^2}} \tag{F2–5}$$

with $n - 2$ degrees of freedom.

ISP COMMANDS

The command SCVCR computes the covariance and correlation of two variables and also has an option for plotting a scatter diagram.

SOLVED EXAMPLE

The data in Table F1.1 provide the number of years of education and the income for $n = 15$ 40-year-olds. The scatter diagram for this data set (Figure F1.1) suggests that the two variables, years of education and income, may be positively related. With the measures introduced here in Chapter F2, we can investigate this relationship.

Let x represent the number of years of education for a 40-year-old and let y represent that person's current income. The sample means and standard deviations for these two variables based on the data in Table F1.1 are

$$\bar{x} = 12.8 \quad , \qquad s_x = 2.426 \quad ,$$
$$\bar{y} = 31{,}820 \quad , \qquad s_y = 5381 \quad .$$

These measures tell us about the average education and income and about the variability in education and income in our sample. However, they do not tell us anything about the relationship between education and income. To investigate the relationship between these variables, we will look at their covariance and correlation.

The easiest way to compute the covariance and correlation of two variables is to use the command SCVCR. Output from SCVCR for the study of the relationship between education and income is given in Figure F2.1. The covariance and correlation coefficient are

$$\text{cov}(x,y) = 8318.59$$

and

$$r = 0.6372 \quad .$$

The positive sign of the covariance and correlation coefficient indicates that the relationship is positive. Higher incomes and more education tend to go together. The magnitude of r suggests that the relationship is by no means perfect, but neither is it an extremely weak relationship. For a test of

$$H_0: \rho = 0$$

versus

$$H_A: \rho \neq 0 \quad ,$$

the t-statistic from (F2-5) is

$$t = \frac{0.6372\sqrt{15 - 2}}{\sqrt{1 - (0.6372)^2}} = 2.981$$

with $15 - 2 = 13$ degrees of freedom. At the $\alpha = 0.05$ level, this is in the rejection region, which consists of $t \leq -2.160$ and $t \geq 2.160$. Thus, we conclude that a sample correlation coefficient of $r = 0.6372$ is highly unlikely to occur by chance in a sample of $n = 15$ if the true population correlation coefficient is zero, and we reject H_0.

FIGURE F2.1   Some output from SCVCR involving the covariance and correlation of years of education and income

```
⟨n⟩ SCVCR
***** Computes the covariance and correlation of 2 variables *****

Need help ? N
Which 2 columns do you want to work on ? 1,2
The covariance of variables     1(EDUC) ,   2 (INCOM) is     8318.5908
The correlation of variables    1(EDUC) ,   2 (INCOM) is     0.6372
```

EXERCISES

F2.1.

Use SCVCR to find the covariance and correlation for the height/weight data in Table F1.2. Can we reject the hypothesis of a correlation of zero between height and weight, with $\alpha = 0.05$?

F2.2.

With the procedure described in Exercise F1.2, enter a data set consisting of the integers from 1 to 20 in ISP and use the transformations TCC, TSQT, TLOG, and TEXP on this set of integers to create new variables. Then use SCVCR to find the correlation of the integers with their squares, with their square roots, with their logarithms, and with an exponential transformation. In each case, have ISP generate a scatter diagram, and see how the values of the correlations seem in relation to the scatter diagrams.

F2.3.

In Exercise F1.3, find the covariance and correlation of the high temperature and the peak load. Test the hypothesis that the correlation is zero against the alternative that it is not zero, using an alpha of 0.01.

F3

Linear Regression

When two variables are related, knowledge about one of the variables can help us predict the other. In simple linear regression, we try to fit a straight line to a set of data on two variables, expressing one variable as a linear function of the other. The variable being predicted is called the dependent variable, and the other variable is called the independent variable. If y is the dependent variable and x the independent variable, the linear regression can be written in the form

$$y = b_0 + b_1 x + e \quad , \tag{F3-1}$$

where e represents an error term which reflects the fact that the relationship between x and y is not perfect. For a given observation, we can determine a predicted value $\hat{y}_i$ based on the value of x_i:

$$\hat{y}_i = b_0 + b_1 x_i \quad . \tag{F3-2}$$

The actual prediction error for this observation is then

$$e_i = y_i - \hat{y}_i \quad . \tag{F3-3}$$

A technique often used in statistics to find an estimated regression line is called the least squares technique. It involves choosing a line so as to minimize the sum of the squared errors, $\Sigma_{i=1}^{n} e_i^2$. The values of b_1 and b_0 that provide a line that best fits the data in a least-squares sense are

$$b_1 = \frac{n\Sigma x_i y_i - (\Sigma x_i)(\Sigma y_i)}{n\Sigma x_i^2 - (\Sigma x_i)^2} \tag{F3-4}$$

and

$$b_0 = \frac{\Sigma y_i - b_1 \Sigma x_i}{n} \quad , \tag{F3-5}$$

where all summations are from $i = 1$ to $i = n$. The slope of the line is b_1, and the y-intercept is b_0. The slope can be related to r, the sample correlation coefficient, as follows:

$$b_1 = r \frac{s_y}{s_x} .$$

(F3–6)

The fitting of a straight line to a set of points on a scatter diagram is descriptive in the sense that it attempts to describe the relationship between x and y in the sample data. In the next chapter, we consider inferences about the relationship between x and y in the population from which the sample has been drawn.

ISP COMMANDS

The ISP command REGRE can be used to calculate b_0 and b_1 and thereby determine an estimated regression line. It also provides much more output, which will be discussed in Chapters F4 and F5.

SOLVED EXAMPLE

High-school grade-point averages are used by some college admissions offices to help predict the performance of students in college. The high-school and college grade-point averages of a sample of 40 students at a large university are shown in Table F3.1. All grade-point averages are expressed on a four-point scale, with 4 (A) being the highest grade.

A scatter diagram of the 40 high-school and college grade-point averages is presented in Figure F3.1. As might be expected, students with high high-school grade-point averages tend also to perform well in college, and those with low high-school grade-point averages are less likely to do well in college. However, there is a considerable amount of variability. Most students know someone who took it easy in high school but suddenly got serious and earned top grades in college, and someone who was a star in high school and never quite got the hang of it in college. These may be exceptions, but they contribute to the variability that makes the relationship between high-school and college performance much less than a perfect relationship.

Some output from REGRE for this example is shown in Figure F3.2. The independent variable x is high-school grade-point average, and the dependent variable y is college grade-point average. Included in the output is the correlation coefficient, which is $r = 0.387$. The estimated regression coefficients are given in the column headed "B". The first coefficient is $b_0 = 1.286$, and the second is $b_1 = 0.544$. The estimated regression line is therefore

$$\hat{y} = 1.286 + 0.544x .$$

TABLE F3.1 High-school and college grade-point averages for a sample of 40 students

Observ.	HSGPA	COGPA
* 1*	3.586	2.951
* 2*	2.969	3.496
* 3*	2.780	2.939
* 4*	3.284	3.078
* 5*	3.509	2.570
* 6*	3.509	3.672
* 7*	3.331	3.523
* 8*	3.297	3.126
* 9*	3.575	3.087
* 10*	3.094	3.500
* 11*	2.921	3.154
* 12*	3.129	2.319
* 13*	3.058	2.715
* 14*	3.844	2.709
* 15*	3.603	3.420
* 16*	3.080	2.720
* 17*	2.760	2.898
* 18*	2.845	2.878
* 19*	2.317	3.126
* 20*	3.530	3.288
* 21*	3.049	3.682
* 22*	3.212	2.836
* 23*	3.025	3.268
* 24*	3.797	3.572
* 25*	2.830	1.858
* 26*	3.269	2.441
* 27*	3.152	2.508
* 28*	2.868	2.419
* 29*	3.270	3.453
* 30*	3.344	2.948
* 31*	3.770	3.476
* 32*	3.091	3.288
* 33*	3.687	3.603
* 34*	3.410	3.171
* 35*	3.021	3.203
* 36*	3.604	3.907
* 37*	2.961	2.577
* 38*	3.173	2.646
* 39*	2.612	2.923
* 40*	3.511	2.992
Sum	129.6768	121.9389
n =	40	40
Mean	3.2419	3.0485

FIGURE F3.1 Scatter diagram of high-school and college grade-point averages for 40 students

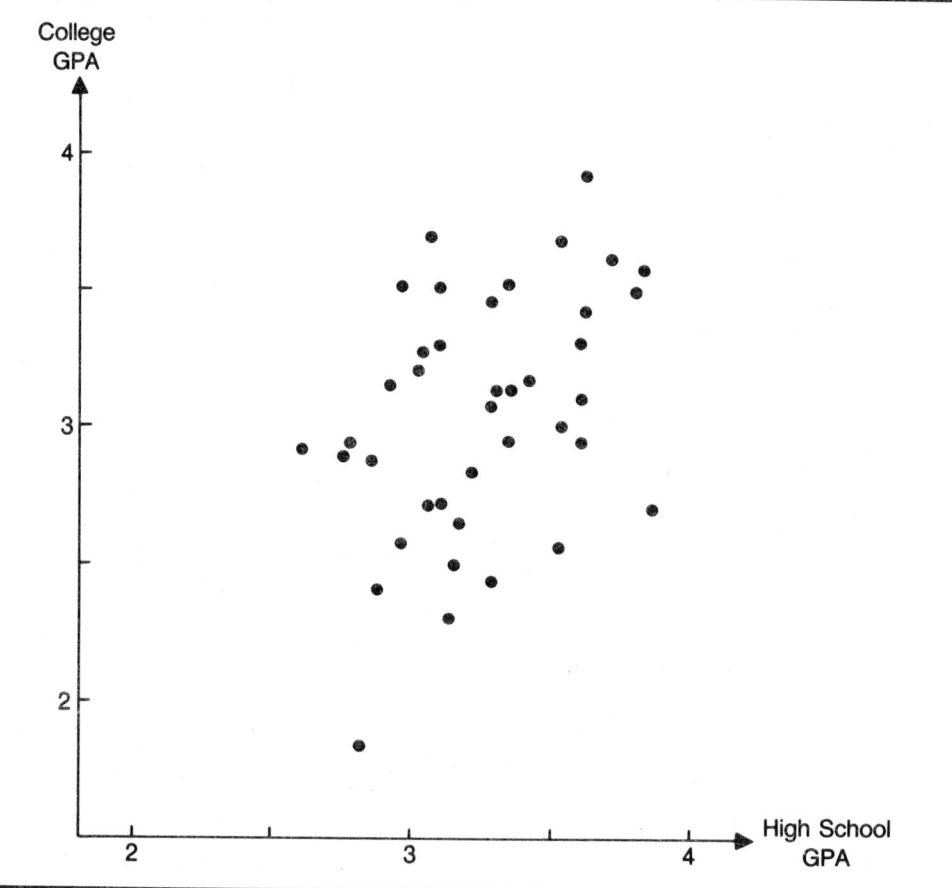

FIGURE F3.2 Some output from REGRE for prediction of college grade-point average as a linear function of high-school grade-point average

⟨n⟩ REGRE

***** Regression analysis *****

Need help ? N
In which columns of your matrix are the independent variables ?
1
In which column of your matrix is your dependent variable ? 6
**** Dependent Variable is :   6(COPGA)
 Independent Variables are :   1

Matrix of Simple Correlation Coefficients

Variab.	HSGPA	COGPA
HSGPA	1.000	0.387
COGPA	0.387	1.000

Variable	B
b (0)	1.286
X(1) = HSGPA	0.544

This estimated regression line is shown on the scatter diagram in Figure F3.3.

The command REGRE produces much more output than is given in Figure F3.2. In the next chapter we will continue to analyze this example, making inferences about the regression line and predictions from the estimated regression line.

EXERCISES

F3.1.

For the data in Table F1.2, use REGRE to find an estimated regression line expressing weight as a linear function of height. Also, use SVAR and SCVCR to find the standard deviations and the correlation coefficient and then use (F3-6) to determine b_1.

F3.2.

For the data in Table F1.3, use REGRE to find an estimated regression line with temperature as the independent variable and peak load as the dependent variable. From this estimated regression line, how much does the peak load change on average when the high temperature increases by one degree? For a high temperature of 92, what is the predicted peak load from the estimated regression line?

FIGURE F3.3   Scatter diagram and estimated regression line for predicting college grade-point average as a function of high-school grade-point average

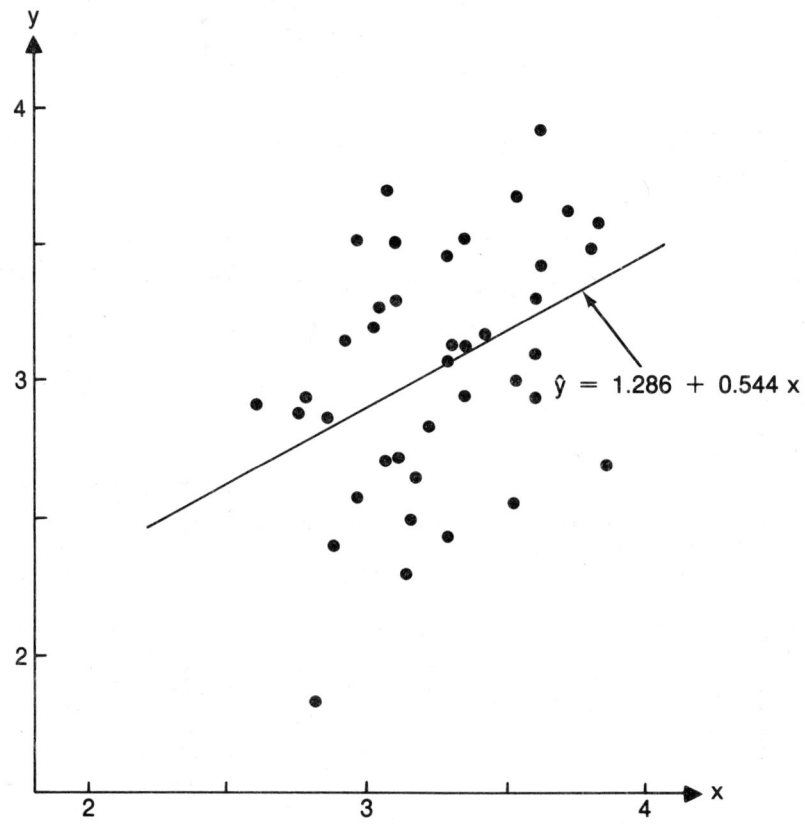

F3.3.

From Exercise F2.2, conduct a regression analysis with the integers as the independent variable and their squares as the dependent variable. Plot the estimated regression line on the scatter diagram and comment on the fit of the linear regression model in this situation.

F4

Inferences in Linear Regression

Suppose that the regression line for the population is denoted by

$$y = \beta_0 + \beta_1 x \quad . \tag{F4-1}$$

Here the Greek letter beta is used with subscripts (β_0 and β_1) to represent the y-intercept and slope of the population regression line. Then the regression line based on the sample,

$$y = b_0 + b_1 x \quad , \tag{F4-2}$$

can be viewed as an estimated regression line. Thus, b_0 provides an estimate of β_0, and b_1 provides an estimate of β_1.

In order to say something about the accuracy of b_0 and b_1 as estimates of β_0 and β_1, we need to make some assumptions. The assumptions are usually expressed in terms of prediction errors. As discussed in Chapter F3, an error is the difference between an actual y-value and the predicted y-value from the regression line. That is, it is the vertical distance between a point on a scatter diagram and the regression line. We assume that these errors are independent of each other and normally distributed with mean zero and variance σ_e^2. Testing the validity of these assumptions in practice will be discussed in Chapter F5.

An estimate of the error variance σ_e^2 is provided by

$$s_e^2 = \frac{\sum_{i=1}^{n} e_i^2}{n - 2} \quad , \tag{F4-3}$$

where

$$e_i = y_i - (b_0 + b_1 x_i) \quad . \tag{F4-4}$$

The square root of s_e^2,

$$s_e = \sqrt{\frac{\sum_{i=1}^{n} e_i^2}{n-2}} \, , \tag{F4–5}$$

is sometimes called the standard deviation of the regression or the standard deviation about the regression line.

For inferences about the regression coefficients β_0 and β_1, we use the t-statistics

$$t = \frac{b_0 - \beta_0}{s_e \, (\sqrt{\sum x_i^2/n})/s_x \sqrt{n-1}} \tag{F4–6}$$

and

$$t = \frac{b_1 - \beta_1}{s_e/s_x \sqrt{n-1}} \, , \tag{F4–7}$$

both of which have $n - 2$ degrees of freedom. The summation in the denominator of (F4–6) and in the following formulas (F4–8) and (F4–10) goes from $i = 1$ to $i = n$. Thus, error bounds with probability $1 - \alpha$ are

$$E = t_{\alpha/2} \frac{s_e \sqrt{\sum x_i^2/n}}{s_x \sqrt{n-1}} \tag{F4–8}$$

when b_0 is used to estimate β_0 and

$$E = t_{\alpha/2} \frac{s_e}{s_x \sqrt{n-1}} \tag{F4–9}$$

when b_1 is used to estimate β_1. The corresponding confidence intervals for β_0 and β_1 are

$$b_0 \pm t_{\alpha/2} \frac{s_e \sqrt{\sum x_i^2/n}}{s_x \sqrt{n-1}} \tag{F4–10}$$

and

$$b_1 \pm t_{\alpha/2} \frac{s_e}{s_x \sqrt{n-1}} \, . \tag{F4–11}$$

Tests concerning β_0 and β_1 are based on (F4–6) and (F4–7), with the value of β_0 from the null hypothesis replacing β_0 in (F4–6) and the value of β_1 from H_0 similarly replacing β_1 in (F4–7).

In general, the proportional reduction in the sum of squared errors achieved by using the estimated regression line is simply R^2, the square of the correlation coefficient. We call R^2 the coefficient of determination, and we use a capital R in R^2 in order to be consistent with the notation for multiple regression in Chapter F7. The goodness of fit of the linear regression can be tested by using the statistic

$$F = \frac{R^2}{(1 - R^2)/(n - 2)} \ ,$$

(F4–12)

which has an F distribution with 1 and $n - 2$ degrees of freedom. The test is one-tailed to the right; if the computed F-statistic is greater than the critical value, we reject the null hypothesis that there is no linear relationship in favor of the alternative that the linear model is useful for predicting the dependent variable.

The ultimate aim of a regression analysis is often the prediction of y-values not included in the sample. The estimated regression line is used for this purpose. The appropriate x-values are plugged into the estimated regression line to generate predictions for y. For $i > n$ (that is, for the prediction of y-values not included in our original sample of size n), the estimated standard deviation of prediction error is

$$s_{e_i} = s_e \sqrt{1 + \frac{1}{n} + \frac{(x_i - \bar{x})^2}{s_x^2 (n - 1)}} \ .$$

(F4–13)

Moreover,

$$t = \frac{\hat{y}_i - y_i}{s_{e_i}}$$

(F4–14)

has a t-distribution with $n - 2$ degrees of freedom. Thus, a confidence interval for y_i with confidence level $1 - \alpha$ is

$$\hat{y}_i \pm t_{\alpha/2}\, s_{e_i} \ .$$

(F4–15)

ISP COMMANDS

The ISP command REGRE does much more than simply compute the estimated regression coefficients b_0 and b_1. It also finds the standard errors of b_0 and b_1, the t-statistics for tests of $\beta_0 = 0$ and $\beta_1 = 0$ (along with a critical t-value); the correlation coefficient; the coefficient of determination; the F-statistic for the test of goodness of fit of the regression line (along with a critical F-value); the standard deviation of the regression; a list of the actual y-values, the predicted y-values, the errors, and the percentage errors; predictions of y for additional observations not in the sample; and other items to be discussed in the next chapter.

SOLVED EXAMPLE

We are now ready to continue our example from Chapter F3 involving the use of linear regression to predict a student's college grade point average as function of the student's high-school grade-point average. The data are given in Table F3.1, and a scatter diagram with the estimated regression line $\hat{y} = 1.286 + 0.544x$ is shown in Figure F3.3. Some output from REGRE is presented in Figure F3.2, but we will focus on the additional output displayed in Figure F4.1.

Next to the estimated regression coefficients $b_0 = 1.286$ and $b_1 = 0.544$ in Figure F4.1 are the standard errors of these coefficients, $s_{b_0} = 0.684$ and $s_{b_1} = 0.210$. Formulas for these standard errors are given in the denominators of (F4–6) and (F4–7). Thus, error bounds with probability 0.95 for the use of b_0 and b_1 as estimates of β_0 and β_1 are

$$E = 1.96\ (0.684) = 1.341$$

for the intercept term and

$$E = 1.96(0.210) = 0.412$$

for the slope of the regression line. The value 1.96 is used for t because with $n - 2 = 40 - 2 = 38$ degrees of freedom, we treat a t-statistic as if it were a normal z-statistic. In terms of 95 percent confidence intervals, we have

$$1.286 \pm 1.341, \text{ or } (-0.055, 2.627)$$

FIGURE F4.1   Some output from REGRE for linear regression for prediction of college grade-point average from high-school grade-point average

Variable	B	Std. error	T-test
$b(\ 0)$	1.286	0.684	1.880
$x(\ 1) =$ HSGPA	0.544	0.210	2.587

The critical t-value from the table (alpha $= 0.05$) $= 1.9600000$
R-squared $= 0.150$ R-squared adjusted $= 0.150$ R $= 0.387$
F-Test $=$ 6.69   Std. dev. of regr. $=$ 0.41
Degrees of freedom for numer. $= 1$, for denomin. $= 38$
F-value from table (alpha $= 0.05$) $= 4.0800000$

for β_0 and

$$0.544 \pm 0.412, \text{ or } (0.132, 0.956)$$

for β_1.

The t-statistics given next to the standard errors of b_0 and b_1 are t-statistics for tests of the hypotheses $\beta_0 = 0$ and $\beta_1 = 0$. They are the statistics given in (F4–6) and (F4–7) with β_0 and β_1 set equal to zero. For the test concerning β_0, $t = 1.88$, which is not in the rejection region for a two-tailed test at significance level 0.05. For the test on β_1, however, the t-value of 2.587 is in the rejection region. Note that the critical value of 1.96 for a two-tailed test at the 0.05 level of significance is included in the REGRE output.

An indication of the overall fit of the linear regression is provided by $R^2 = 0.150$. (The adjusted R^2 in the REGRE output is used only in multiple regression and can be ignored in linear regression with only one independent variable.) The F-statistic from (F4–12) is 6.69 with 1 and 38 degrees of freedom, as compared with a critical value of $F = 4.08$ at the 0.05 level of significance. Thus, we conclude that a linear relationship is present, although the estimated linear regression results in only a 15% reduction in the sum of squared prediction errors. A final piece of information provided in Figure F4.1 is the standard deviation of the regression, $s_e = 0.41$.

The command REGRE will also generate predictions from the estimated regression equation. Suppose that we have five more students (not included in the original sample of 40 students). Their high-school grade-point averages are $x_{41} = 2.6$, $x_{42} = 2.9$, $x_{43} = 3.2$, $x_{44} = 3.5$, and $x_{45} = 3.8$. As shown in Figure F4.2, the college grade-point averages predicted for these students from the estimated regression line are

$$\hat{y}_{41} = 2.700 \quad,$$
$$\hat{y}_{42} = 2.863 \quad,$$
$$\hat{y}_{43} = 3.026 \quad,$$
$$\hat{y}_{44} = 3.189 \quad,$$
$$\text{and } \hat{y}_{45} = 3.352 \quad.$$

FIGURE F4.2 Some output from REGRE for the prediction of college grade-point averages for five students from an estimated regression line

Do you want to forecast with your regression model ? Y
How many forecasts do you want ? 5
** Enter below the values for each of your   1 independent variables **
Enter for observation   41
The value for the independent variable :   1
2.6
Enter for observation   42
The value for the independent variable :   1
2.9
Enter for observation   43
The value for the independent variable :   1
3.2
Enter for observation   44
The value for the independent variable :   1
3.5
Enter for observation   45
The value for the independent variable :   1
3.8

Observation	Forecast
41	2.700
42	2.863
43	3.026
44	3.189
45	3.352

EXERCISES

F4.1.

As in Exercise F3.1, use REGRE to find an estimated regression line expressing weight as a linear function of height. Find the standard errors of b_0 and b_1 and the t-statistic for the test of the null hypothesis that b_1 is zero. What is the proportional reduction in the sum of squared errors achieved by using the estimated regression line? Test for the overall goodness of fit of the regression line. Taking all of the output from REGRE into account, discuss this regression analysis.

F4.2.

For the data in Table F1.3, use REGRE to find an estimated regression line with high temperature as the independent variable and peak load as the dependent variable. Find 95 percent confidence intervals for β_0 and β_1. If we have a day (not included in the sample) with a high temperature of 93°, find the peak load predicted from the estimated regression line, the standard deviation of the prediction error, and a 95 percent interval for the peak load on this day.

F4.3.

The data in Table F4.1 show weekly sales of an item at different prices. Use REGRE to conduct a regression analysis expressing sales as a linear function of price. Discuss the implications of the estimated coefficients b_0 and b_1 in terms of the sales/price relationship. Check out the various inferences discussed in this chapter and comment on the regression analysis in view of these results. Then, for prices of 12 and 15, use the estimated regression line to predict sales and indicate how uncertain we should be about the predictions.

TABLE F4.1   Weekly sales of an item at different prices

Sales	Price	Sales	Price	Sales	Price
47	11	36	17	38	14
42	15	58	12	32	20
47	13	38	16	61	13
31	19	35	18	46	14

F5
Testing the Assumptions of Regression

Checking on the assumptions underlying a statistical procedure is an important step in any statistical analysis because serious violations of the assumptions may render our inferences invalid. In regression analysis, we assume that the errors are independent and normally distributed with mean zero and constant variance. Thus, an investigation of the assumptions of regression focuses on the errors, which are often called residuals, found from (F3–3). An analysis of the residuals includes informal graphical procedures as well as some formal statistical tests.

Various graphs involving the residuals can shed light on possible violations of the assumptions. For example, the residuals can be plotted in the order of the observations or plotted against the values of the independent variable x. The existence of systematic patterns (for example, if the positive residuals tend to bunch together and the negative residuals tend to bunch together) suggests problems with the assumptions.

Formal tests can also be used. The assumption of independence of the errors can be tested by using a Durbin-Watson test, which is based on the statistic

$$D = \frac{\sum_{i=2}^{n} (e_i - e_{i-1})^2}{\sum_{i=1}^{n} e_i^2} \ .$$ (F5–1)

This statistic is compared with values d_L and d_U from a table, and the decision rule is as follows:

Accept the hypothesis of independence if $d_U < D < 4 - d_U$.
The test is inconclusive if $d_L < D \leq d_U$ or $4 - d_U \leq D < 4 - d_L$.
Reject the hypothesis of independence if $D \leq d_L$ or $D \geq 4 - d_L$.

The assumption of normality of the errors can be investigated by looking at a histogram of the residuals and using a chi-square goodness of fit test to see whether a normal curve provides a good fit to this histogram. If the informal graphical procedures or formal tests indicate problems, we may need to modify our regression analysis, perhaps by considering nonlinear models (as in Chapter F6) or including additional variables (as in Chapter F7).

ISP COMMANDS

The ISP command REGRE, in addition to the output described in Chapters F3 and F4, provides a list of residuals; a list of residuals in percentage error form; and the computed Durbin-Watson statistic along with values of d_L and d_U and the decision rule for the Durbin-Watson test. Moreover, the residuals (and the predicted y-values) can be moved to the data matrix and stored there for further computation or graphing. The command PLOT can then be used to plot the residuals against the observation number, and PSCAT can be used to plot the residuals against x or y or even the predicted y-values. Of course, PHIST can provide a histogram of the residuals and PDIST can give a chi-square goodness of fit test to check the normality of the residuals.

SOLVED EXAMPLE

It is well known that alcoholic drinks can reduce a person's ability to perform certain tasks that require concentration or agility. Here we consider an experiment in which subjects drink varying numbers (0–6) of alcoholic drinks and then try to perform a set routine of tasks. The number of drinks a person has and the number of mistakes on the tasks are recorded. The data for 28 experimental subjects are shown in Table F5.1.

Linear regression can be used to attempt to predict the number of mistakes, y, as a function of the number of drinks, x. Some output from REGRE is shown in Figure F5.1. First, note that the inferential results seem very impressive. The correlation coefficient is 0.914 and the coefficient of determination is 0.836, indicating that the estimated linear regression reduces the sum of squared prediction errors by 83.6 percent. The F-statistic of 132.44 for a test of the goodness of fit of the linear model is much higher than the critical F-value. The estimated regression line is

$$\hat{y} = 2.152 + 3.116x \quad,$$

and the standard errors of b_0 and b_1 are 0.976 and 0.271, respectively. Thus, 95 percent confidence intervals are

$$2.152 \pm 2.056 \, (0.976), \text{ or } (0.145, 4.159)$$

for β_0 and

$$3.116 \pm 2.056 \, (0.271), \text{ or } (2.559, 3.673)$$

TABLE F5.1 Number of drinks and number of mistakes for 28 experimental subjects

Observ.		Drink	Mis
*	1*	0.000	7.000
*	2*	0.000	4.000
*	3*	0.000	7.000
*	4*	0.000	6.000
*	5*	1.000	5.000
*	6*	1.000	5.000
*	7*	1.000	3.000
*	8*	1.000	4.000
*	9*	2.000	4.000
*	10*	2.000	7.000
*	11*	2.000	7.000
*	12*	2.000	8.000
*	13*	3.000	11.000
*	14*	3.000	9.000
*	15*	3.000	10.000
*	16*	3.000	9.000
*	17*	4.000	10.000
*	18*	4.000	14.000
*	19*	4.000	12.000
*	20*	4.000	13.000
*	21*	5.000	17.000
*	22*	5.000	17.000
*	23*	5.000	17.000
*	24*	5.000	18.000
*	25*	6.000	26.000
*	26*	6.000	22.000
*	27*	6.000	23.000
*	28*	6.000	27.000
Sum		84.0000	322.0000
n =		28	28
Mean		3.0000	11.5000

for β_1. The t-statistic for the test of H_0: $\beta_1 = 0$ is 11.508, which is strong evidence that β_1 is not zero. The t-statistic for the test of H_0: $\beta_0 = 0$ is 2.204, which is in the rejection region at the 0.05 level of significance.

Are all of these inferences valid? The list of residuals is given in Figure F5.1 and plotted against the observation number in Figure F5.2. We can see a systematic pattern in which the first four and last five residuals are positive and the remaining residuals are negative. The value of D for the Durbin-Watson statistic is 0.71, which is in the rejection region. It appears that the residuals are not independent.

In an experiment with no obvious time effect, why would the residuals be positively correlated? Note, from Table F5.1, that the observations have been recorded in order of

FIGURE F5.1   Output from REGRE for prediction of number of mistakes as a function of number of drinks

Variable	B	Std. error	T-test
b(0)	2.152	0.976	2.204
X(1) = DRINK	3.116	0.271	11.508

The critical t-value from the table (alpha = 0.05) = 2.0560000

R-Squared = 0.836 R-Squared adjusted = 0.836 R = 0.914
  F-test =   132.44   Std. dev. of regr. =   2.87
Degrees of freedom for numer. = 1 , for denomin. = 26

F-value from table (alpha = 0.05) = 4.2400000
Do you want a list of the actual values and forecasts ? Y

	Actual	Forecast	Residuals	% Error
1	7.0000000	2.1517859	4.8482141	69.260201
2	4.0000000	2.1517859	1.8482141	46.205353
3	7.0000000	2.1517859	4.8482141	69.260201
4	6.0000000	2.1517859	3.8482141	64.136902
5	5.0000000	5.2678576	− 0.26785731	− 5.3571463
6	5.0000000	5.2678576	− 0.26785731	− 5.3571463
7	3.0000000	5.2678576	− 2.2678573	− 75.595245
8	4.0000000	5.2678576	− 1.2678573	− 31.696432
9	4.0000000	8.3839293	− 4.3839288	− 109.59822
10	7.0000000	8.3839293	− 1.3839288	− 19.770411
11	7.0000000	8.3839293	− 1.3839288	− 19.770411
12	8.0000000	8.3839293	− 0.38392878	− 4.7991099
13	11.000000	11.500000	− 0.50000000	− 4.5454545
14	9.0000000	11.500000	− 2.5000000	− 27.777779
15	10.000000	11.500000	− 1.5000000	− 15.000001
16	9.0000000	11.500000	− 2.5000000	− 27.777779
17	10.000000	14.616072	− 4.6160717	− 46.160717
18	14.000000	14.616072	− 0.61607170	− 4.4005122
19	12.000000	14.616072	− 2.6160717	− 21.800596
20	13.000000	14.616072	− 1.6160717	− 12.431320
21	17.000000	17.732143	− 0.73214340	− 4.3067260
22	17.000000	17.732143	− 0.73214340	− 4.3067260
23	17.000000	17.732143	− 0.73214340	− 4.3067260
24	18.000000	17.732143	0.26785660	1.4880922
25	26.000000	20.848213	5.1517868	19.814564
26	22.000000	20.848213	1.1517868	5.2353945
27	23.000000	20.848213	2.1517868	9.3555946
28	27.000000	20.848213	6.1517868	22.784395

Durbin-Watson test calculated =   0.71
Durbin-Watson test from table (for   1 indep. variables and   28 obs.):
  Upper =   1.48, Region A (Accept) : 1.48 to 2.52
  Lower =   1.33, Region B (?)   : 1.33 to 1.48 and 2.52 to 2.67
                       Region C (Reject) : 0   to 1.33 and 2.67 to 4.00

FIGURE F5.2   Plot of residuals against observation number for prediction of number of mistakes as a function of number of drinks

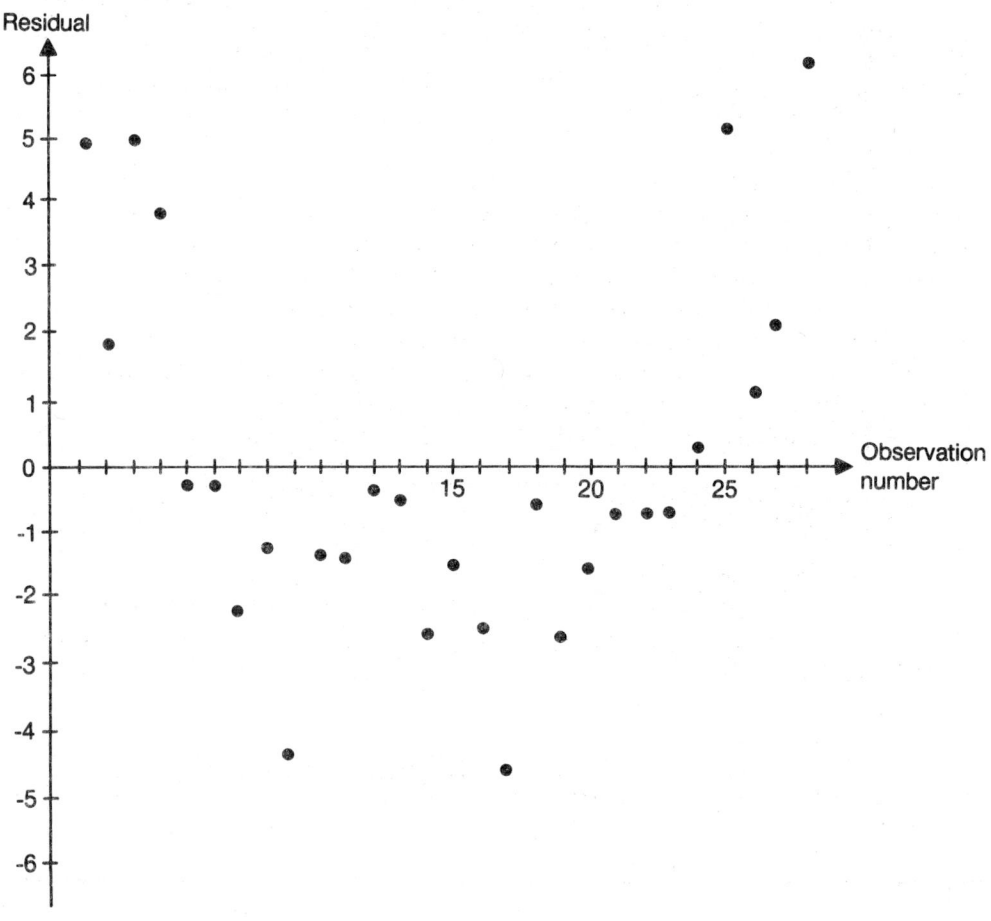

number of drinks, with the subjects having no drinks listed first, the subjects having one drink listed next, and so on. Thus, the non-independence of residuals may reflect a relationship between the residuals and the independent variable, number of drinks. The residuals are plotted against the number of drinks in Figure F5.3, and we see a definite pattern in this graph. This indicates that a nonlinear function is more suitable than a linear function for this regression problem.

This example illustrates a very important point. Even when the inferential results (R^2, the F-test, the t-tests for individual coefficients β_0 and β_1) look very good, it is important to check the regression assumptions. If the assumptions are highly inappropriate,

FIGURE F5.3   Plot of residuals against number of drinks for prediction of number of mistakes as a function of number of drinks

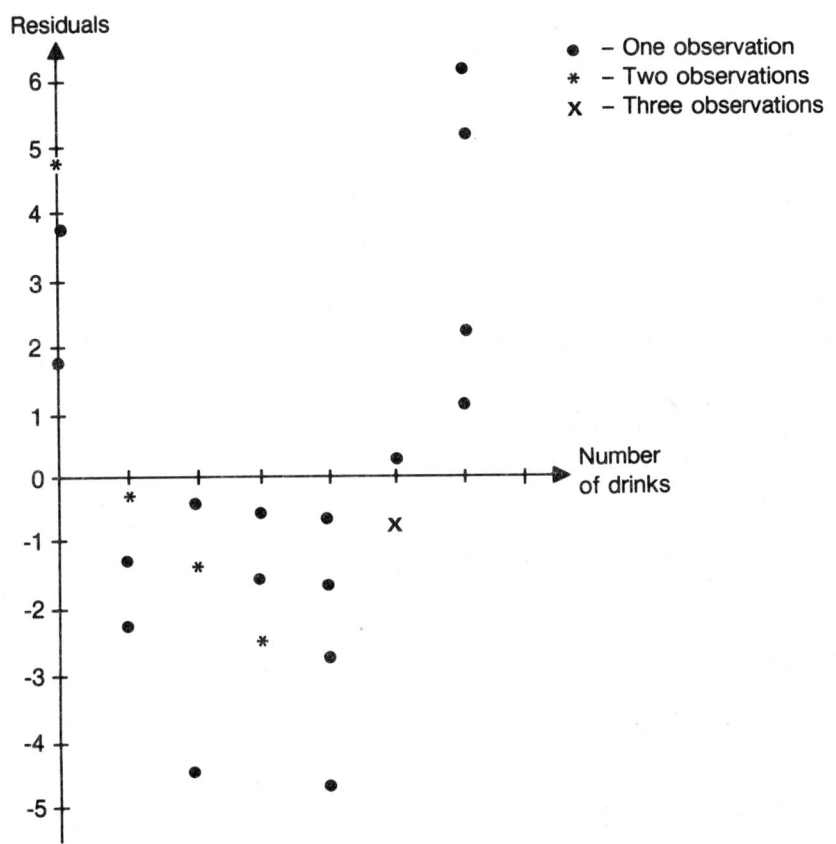

the inferences may be meaningless and a different analysis (such as a nonlinear regression or multiple regression) may be much more suitable.

EXERCISES

F5.1.

For the estimated regression from Exercise F4.1 with weight as a function of height, save the residuals from REGRE. Use PLOT to plot the residuals in order, and use PSCAT to plot the residuals against weight and against height. Finally, use PHIST to plot a histogram of the residuals. Do these graphs suggest any violations of the regression assumptions?

F5.2.

Conduct an analysis of the residuals for the regression with peak load as a function of high temperature (Exercise F4.2). Include graphical analyses and the Durbin-Watson test. What do you conclude about the regression assumptions in this example?

F5.3.

From Exercise F2.2, conduct a regression analysis with the integers as the independent variable and their squares as the dependent variable. Analyze the residuals and discuss.

F6
Transformations and Nonlinear Relationships

Some pairs of variables happen to be related in a nonlinear fashion. The consideration of nonlinear relationships through the use of transformations allows us to retain the general linear regression approach and its relative ease of use. Instead of being linear in x, our function may be linear in terms of some transformed variable such as x^2, log x, or e^x. Transformations can also be used on the dependent variable. For example, if the variance of the error term appears to increase as x increases, transforming y by taking its square root may help to stabilize the variance. The possibility of using transformations greatly increases the flexibility of regression analysis. Once the transformation has been made, the mechanics (determining least squares estimates, checking the assumptions of regression, making inferences and predictions) are exactly as described in the preceding chapters.

ISP COMMANDS

Various T commands are available for performing transformations on variables. TCC squares a variable, TSQT takes the square root, TLOG provides the natural logarithm of a variable, TPOW raises a variable to a power, and TEXP transforms x to e^x. Other transformations can be achieved by using combinations of commands. For example, using TCON to create a column of ones and then using TDIV to divide this column by the column for a variable gives the reciprocal of the variable. Of course, PSCAT can be used to plot scatter diagrams, which can be useful in the decision as to whether to consider a transformation and if so, which transformation. Once the transformation has been made, the regression analysis with the transformed variables can be provided by REGRE with assistance from PLOT, PHIST, and PSCAT as noted in Chapter F5.

SOLVED EXAMPLE

Manufacturers of luxury cars often claim that their cars maintain their value to a greater extent than do cheaper cars. Suppose that we take a sample of 24 cars of a particular manufacturer. We sample 3 one-year-old cars, 3 two-year-old cars, and so on up to 3 eight-year-old cars. All of the 24 cars in the sample sold recently and the prices are given along with the ages in Table F6.1.

A scatter diagram of the 24 observations of age-price combinations is shown in Figure F6.1. A logarithmic function (with a negative coefficient for the logarithmic term) might provide a good fit to this data. Therefore, we use a logarithmic transformation, calculating the natural logarithm of the age for each car in the sample. The logarithms of the ages are shown in the last column of Table F6.2.

Our regression model is now linear in log x:

$$y = b_0 + b_1 \log x \quad .$$

Output from REGRE is presented in Figure F6.2. The estimated regression equation is

$$\hat{y} = 20{,}413.79 - 7186.54 \log x \quad .$$

TABLE F6.1   The age and selling price of 24 cars

Observ.		Age	Price
*	1*	1.000	19900.000
*	2*	1.000	20100.000
*	3*	1.000	21000.000
*	4*	2.000	15300.000
*	5*	2.000	15500.000
*	6*	2.000	17000.000
*	7*	3.000	12300.000
*	8*	3.000	12200.000
*	9*	3.000	11700.000
*	10*	4.000	10400.000
*	11*	4.000	10300.000
*	12*	4.000	10800.000
*	13*	5.000	8100.000
*	14*	5.000	8500.000
*	15*	5.000	8800.000
*	16*	6.000	7100.000
*	17*	6.000	7200.000
*	18*	6.000	8900.000
*	19*	7.000	6800.000
*	20*	7.000	6600.000
*	21*	7.000	6000.000
*	22*	8.000	5500.000
*	23*	8.000	5200.000
*	24*	8.000	6100.000
Sum		108.0000	261300.0000
n =		24	24
Mean		4.5000	10887.5000

FIGURE F6.1   Scatter diagram of price of a car (y) versus age in years (x)

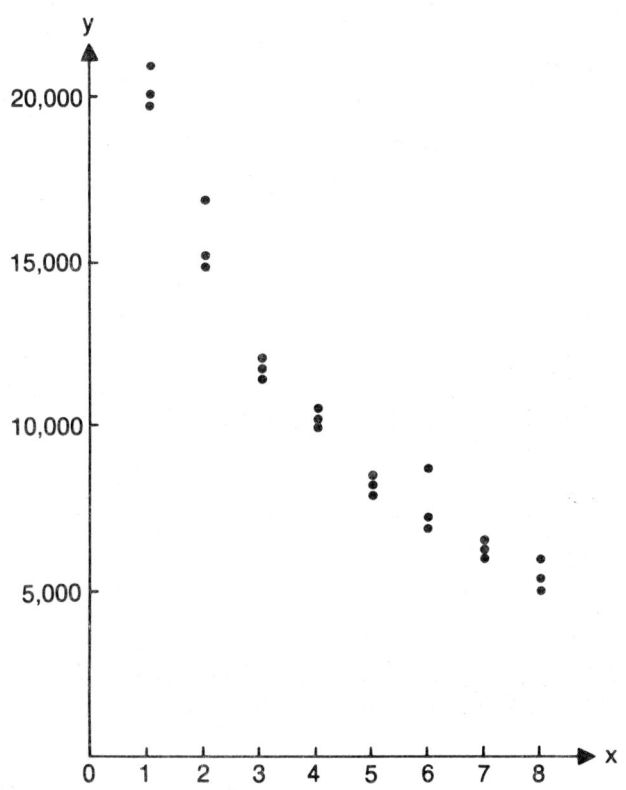

Both regression coefficients are significantly different from zero, and the F-statistic is extremely high. The value of R^2 is 0.986, which is very high. Furthermore, a look at the residuals and the Durbin-Watson test indicates that the regression assumptions are met. The logarithmic transformation results in an excellent model in this case. This can be confirmed visually by a look at Figure F6.3, where the estimated regression equation is graphed on a scatter diagram of y versus x.

Since we are happy with our model, we can use it to forecast. For cars 2, 6, and 8 years old, the estimated prices are, respectively,

$$20{,}413.79 - 7186.54 \log 2 = 15{,}434 \quad ,$$
$$20{,}413.79 - 7186.54 \log 6 = 7536 \quad ,$$
$$\text{and } 20{,}413.79 - 7186.54 \log 8 = 5473 \quad .$$

TABLE F6.2 The age, selling price, and logarithm of age of 24 cars

Observ.	Age	Price	Logag
* 1*	1.000	19900.000	0.000
* 2*	1.000	20100.000	0.000
* 3*	1.000	21000.000	0.000
* 4*	2.000	15300.000	0.693
* 5*	2.000	15500.000	0.693
* 6*	2.000	17000.000	0.693
* 7*	3.000	12300.000	1.099
* 8*	3.000	12200.000	1.099
* 9*	3.000	11700.000	1.099
* 10*	4.000	10400.000	1.386
* 11*	4.000	10300.000	1.386
* 12*	4.000	10800.000	1.386
* 13*	5.000	8100.000	1.609
* 14*	5.000	8500.000	1.609
* 15*	5.000	8800.000	1.609
* 16*	6.000	7100.000	1.792
* 17*	6.000	7200.000	1.792
* 18*	6.000	8900.000	1.792
* 19*	7.000	6800.000	1.946
* 20*	7.000	6600.000	1.946
* 21*	7.000	6000.000	1.946
* 22*	8.000	5500.000	2.079
* 23*	8.000	5200.000	2.079
* 24*	8.000	6100.000	2.079
Sum	108.0000	261300.0000	31.8138
n =	24	24	24
Mean	4.5000	10887.5000	1.3256

As you can see, once we have made the transformation, we treat the transformed variable just like any other variable that is being used in a simple linear regression analysis.

FIGURE F6.2 Some output from REGRE for the prediction of the price of a car in dollars as a function of the natural logarithm of its age in years

Variable	B	Std. error	T-test
b(0)	20413.793	273.695	74.586
X(3) = LOGAG	− 7186.535	184.943	− 38.858

The critical t-value from the table (alpha = 0.05) =   2.0740000

R-Squared = 0.986 R-Squared adjusted = 0.986 R = 0.993
  F-Test =   1509.96   Std. dev. of regr. =   596.14
Degrees of freedom for numer. = 1 , for denomin. = 22

F-value from table (alpha = 0.05) = 4.300000
Do you want a list of the actual values and forecasts ? Y

	Actual	Forecast	Residuals	% Error
1	19900.000	20413.793	− 513.79297	− 2.5818744
2	20100.000	20413.793	− 313.79297	− 1.5611591
3	21000.000	20413.793	586.20703	2.7914622
4	15300.000	15432.467	− 132.46680	− 0.86579603
5	15500.000	15432.467	67.533203	0.43569809
6	17000.000	15432.467	1567.5332	9.2207832
7	12300.000	12518.578	− 218.57764	− 1.7770540
8	12200.000	12518.578	− 318.57764	− 2.6112921
9	11700.000	12518.578	− 818.57764	− 6.9963903
10	10400.000	10451.141	− 51.140625	− 0.49173677
11	10300.000	10451.141	− 151.14063	− 1.4673847
12	10800.000	10451.141	348.85938	3.2301795
13	8100.0000	8847.5117	− 747.51172	− 9.2285395
14	8500.0000	8847.5117	− 347.51172	− 4.0883732
15	8800.0000	8847.5117	− 47.511719	− 0.53990591
16	7100.0000	7537.2510	− 437.25098	− 6.1584644
17	7200.0000	7537.2510	− 337.25098	− 4.6840415
18	8900.0000	7537.2510	1362.7490	15.311787
19	6800.0000	6429.4424	370.55762	5.4493766
20	6600.0000	6429.4424	170.55762	2.5842063
21	6000.0000	6429.4424	− 429.44238	− 7.1573725
22	5500.0000	5469.8145	30.185547	0.54882813
23	5200.0000	5469.8145	− 269.81445	− 5.1887393
24	6100.0000	5469.8145	630.18555	10.330911

Durbin-Watson test calculated =   1.91
Durbin-Watson test from table (for 1 indep. variables and 24 obs.):
   Upper =   1.45,   Region A (Accept) : 1.45 to 2.55
   Lower =   1.27,   Region B (?)   : 1.27 to 1.45 and 2.55 to 2.73
                      Region C (Reject) : 0   to 1.27 and 2.73 to 4.00

FIGURE F6.3 The estimated regression equation $\hat{y} = 20{,}413.79 - 7186.54 \log x$ on the scatter diagram of y (price of a car) versus x (age of the car in years)

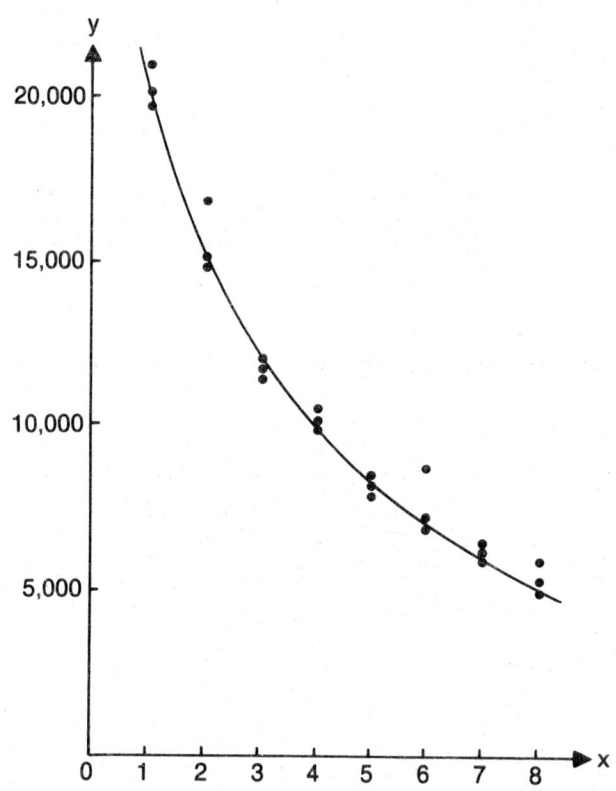

EXERCISES

F6.1.

A manufacturer trying to estimate a cost curve keeps track of the production cost for ten different orders of differing quantities. The data are given in Table F6.3, where cost is expressed in dollars and quantity in thousands of units. Use PSCAT to plot a scatter diagram of cost as a function of quantity, and use REGRE to determine an estimated regression line with cost as a function of quantity. Then use TSQT to transform from quantity to the square root of quantity, and generate a new scatter diagram and regression

TABLE F6.3   Cost (in dollars) and quantity (in thousands) for ten orders

Quantity	Cost
3	740
6	980
4	840
12	1400
3	700
8	1170
1	410
18	1710
7	1130
5	930

analysis with cost as the dependent variable and the square root of quantity as the independent variable. Compare the two scatter diagrams and the two regression analyses, and discuss.

F6.2.

A measure of the level of a pollutant in the air in a given city and the number of heart attacks in the city are shown for 10 days in Table F6.4. Conduct a regression analysis with the logarithm of the level of the pollutant (use TLOG) as the independent variable and the number of heart attacks as the dependent variable.

Table F6.4   Level of a pollutant and number of heart attacks for 10 days

Level of pollutant:	1.0	1.4	2.5	0.8	1.1	2.7	1.3	6.1	4.3	2.3
Number of heart attacks:	5	8	12	4	7	15	6	23	19	14

F7
Multiple Regression

Let the dependent variable be denoted by y, as before, and suppose that there are k independent variables $x_1, x_2, \ldots, x_k$. Note that x_i now refers not to the ith observation on variable x but, instead, to the ith variable. A second subscript can be used to represent the observation number. Thus, x_{24} stands for the fourth observation on x_2, the second variable. Each sample observation now consists not just of a single value or a pair, but of a set of $k + 1$ values, one value for each of the $k + 1$ variables (k independent variables and one dependent variable). For example, the first observation consists of

$$y_1, x_{11}, x_{21}, \ldots, x_{k1} \quad ,$$

the second observation is

$$y_2, x_{12}, x_{22}, \ldots, x_{k2},$$

and so on, with the last of the n observations consisting of

$$y_n, x_{1n}, x_{2n}, \ldots, x_{kn}.$$

The population multiple regression equation is

$$y = \beta_0 + \beta_1 x_1 + \beta_2 x_2 + \ldots + \beta_k x_k \quad , \tag{F7--1}$$

with an intercept term β_0 and k coefficients $\beta_1, \beta_2, \ldots, \beta_k$ for the k independent variables. As in the case of simple linear regression with one independent variable, we want to use the sample data to determine an estimated regression equation,

$$y = b_0 + b_1 + b_2 x_2 + \ldots + b_k x_k \quad . \tag{F7--2}$$

For a given observation i, the amount the actual value y_i differs from the predicted value

$$\hat{y}_i = b_0 + b_1 x_{1i} + b_2 x_{2i} + \ldots + b_k x_{ki} \tag{F7--3}$$

is

$$e_i = y_i - \hat{y}_i \quad .$$

The least squares technique is used to find the estimated regression equation. In this technique, the values of b_0, b_1, b_2,...,b_k are chosen to minimize $\Sigma_{i=1}^{n} e_i^2$. These values cannot be expressed in terms of easy-to-calculate formulas. For multiple regression, computations are virtually always done by a computer.

The interpretation of the output from a multiple regression analysis is very similar to (and in many senses identical to) the interpretation of the output from a simple regression analysis (a regression analysis with only one independent variable). Confidence intervals or tests can be considered for each of the regression coefficients, the assumptions can be checked with an analysis of the residuals, and predictions can be made from the regression equation. In testing the overall fit of the estimated linear regression equation, the multiple R^2 indicates the proportional reduction in the sum of squared prediction errors, and a formal test for the overall fit is provided by the test statistic

$$F = \frac{R^2/k}{(1 - R^2)/(n - k - 1)} \quad , \tag{F7–4}$$

which has an F distribution with k and $n - k$ degrees of freedom. By looking at tests for individual coefficients, analyzing the residuals, and considering the overall fit, we can compare different regression models. This enables us to consider transformations and the possibility of adding variables to or deleting variables from the regression model.

ISP COMMANDS

The command REGRE handles multiple regression as easily as simple regression, and the other commands mentioned in Chapters F3 to F6 may also be helpful in a multiple regression analysis.

SOLVED EXAMPLE

High-school grade-point averages are used by some college admissions offices to help predict the performance of students in college, as illustrated by an example in Chapters F3 and F4. However, that example demonstrates that while the use of high-school grade-point average as the only independent variable improves our predictions of college grade-point average, the improvement is less than overwhelming. In fact, college admissions offices generally consider several factors in attempting to predict college performance. The data in Table F7.1 provide the college and high-school grade-point averages, the SAT mathematics score, and a rating based on a personal interview conducted during the

TABLE F7.1   College grade-point average, high-school grade-point average, SAT mathematics score, and interview rating for a sample of 25 students

OBSERV.	COGPA	HSGPA	SATM	INTRA
* 1*	2.840	3.450	601.000	6.400
* 2*	3.050	3.400	600.000	4.700
* 3*	2.860	3.090	550.000	6.100
* 4*	2.400	3.150	571.000	4.300
* 5*	3.550	3.180	708.000	5.400
* 6*	3.090	3.260	628.000	7.200
* 7*	3.780	3.470	604.000	6.000
* 8*	2.580	3.180	612.000	6.700
* 9*	2.420	3.870	572.000	1.200
* 10*	3.280	3.680	622.000	6.700
* 11*	3.140	2.850	591.000	4.800
* 12*	3.190	3.680	669.000	1.800
* 13*	3.410	3.690	699.000	5.000
* 14*	3.350	3.340	601.000	6.600
* 15*	2.910	2.840	517.000	6.700
* 16*	2.520	3.010	600.000	3.200
* 17*	2.700	3.070	559.000	3.800
* 18*	2.690	2.720	624.000	3.600
* 19*	3.470	3.380	694.000	6.900
* 20*	3.460	3.400	560.000	5.700
* 21*	3.110	3.360	662.000	4.900
* 22*	2.820	3.170	575.000	4.700
* 23*	3.080	3.060	688.000	5.300
* 24*	3.520	2.940	626.000	5.900
* 25*	3.430	3.190	538.000	6.200
Sum	76.6499	81.4300	15271.0000	129.8000
n =	25	25	25	25
Mean	3.0660	3.2572	610.8400	5.1920

student's senior year in high school. A student's high-school grade-point average reflects performance over his or her high-school career, the score on the mathematics portion of the SAT test reflects ability in mathematics (and perhaps in test-taking), and an interview rating may capture some intangibles (extracurricular activities, personality, interests and ambitions, and so on) in addition to academic performance. The interview rating, incidentally, is an average of ratings given by several interviewers, with 8 representing the highest possible rating and 0 the lowest possible rating.

A multiple regression predicting college grade-point average as a function of the other three variables from Table F7.1 is summarized in Figure F7.1. Note that $R^2 = 0.403$ and the F test is significant at the 0.05 level ($F = 4.72$, as compared with a critical value of 3.07). However, of the regression coefficients, only one (the coefficient for the interview rating) has a t-statistic large enough to conclude at the 0.05 level that the

FIGURE F7.1 Some output from REGRE to predict college grade-point average as a function of high-school grade-point average, SAT mathematics score, and interview rating

⟨n⟩ REGRE

***** Regression analysis *****

Need help ? N
In which columns of your matrix are the independent variables ?
2,3,4
In which column of your matrix is your dependent variable ? 1
**** Dependent variable is : 1(COGPA)
  Independent Variables are : 2 3 4

Matrix of Simple Correlation Coefficients

Variab.	HSGPA	SATM	INTRA	COGPA
HSGPA	1.000	0.259	−0.180	0.201
SATM	0.259	1.000	−0.014	0.386
INTRA	−0.180	−0.014	1.000	0.460
COGPA	0.201	0.386	0.460	1.000

Variable	B	Std. error	T-test
b (0)	0.022	0.995	0.022
X(2) = HSGPA	0.268	0.235	1.141
X(3) = SATM	0.003	0.001	1.950
X(4) = INTRA	0.124	0.042	2.922

The critical t-value from the table (alpha = 0.05) =   2.0800000

R-Squared = 0.403 R-Squared adjusted = 0.348 R = 0.635
  F-Test =   4.72   Std. dev. of regr. =   0.32
Degrees of freedom for numer. = 3 , for denomin. = 21

F-value from table (alpha = 0.05) = 3.0700000
Do you want a list of the actual values and forecasts ? N

Durbin-Watson test calculated =   2.15
Durbin-Watson test from table (for 3 indep. variables and 25 obs.):
  Upper = 1.66, Region A (Accept) : 1.66 to 2.34
  Lower = 1.12, Region B (?)   : 1.12 to 1.66 and 2.34 to 2.88
              Region C (Reject) : 0   to 1.12 and 2.88 to 4.00

coefficient is significantly different from zero. This casts doubt on the reliability of the estimated regression equation, which is

$$\hat{y} = 0.022 + 0.268x_1 + 0.003x_2 + 0.124x_3 \quad ,$$

where y represents college grade-point average and the three independent variables are x_1 = high-school grade-point average, x_2 = SAT mathematics score, and x_3 = interview rating.

The matrix of correlation coefficients tells us that the correlation of y with x_1 is 0.201, the correlation of y with x_2 is 0.386, and the correlation of y with x_3 is 0.460. Thus, x_1 has the smallest correlation with y. Also, the t-statistic for the test of $\beta_1 = 0$ is smaller than the t-statistics involving β_2 and β_3. What would happen if we dropped x_1 and just used x_2 and x_3 as independent variables? We can find out quickly by trying this regression using REGRE, and some output is presented in Figure F7.2.

In this regression with only two independent variables, the t-statistics for the coefficients of the two variables are both significant (greater than the critical value) at the 0.05 level of significance. Also, the value of F has increased. The value of R^2 has decreased, as it always does when a variable is eliminated from consideration, but the change is not too great. In fact, the change in the adjusted R^2, which is a measure that adjusts R^2 to allow for the number of variables in the equation, is extremely small. An analysis of residuals indicates that the assumptions of the regression model are satisfied.

The estimated multiple regression equation from Figure F7.2 is

$$\hat{y} = 0.709 + 0.003x_2 + 0.115x_3 \quad .$$

FIGURE F7.2 Some output from REGRE to predict college grade-point average as a function of SAT mathematics score and interview rating

Matrix of Simple Correlation Coefficients

Variab.	SATM	INTRA	COGPA
SATM	1.000	−0.014	0.386
INTRA	−0.014	1.000	0.460
COGPA	0.386	0.460	1.000

Variable	B	Std. error	T-test
b (0)	0.709	0.798	0.889
X(3) = SATM	0.003	0.001	2.312
X(4) = INTRA	0.115	0.042	2.742

The critical t-value from the table (alpha = 0.05) = 2.0740000

R-Squared = 0.366 R-Squared adjusted = 0.338 R = 0.605
  F-Test =   6.34   Std. dev. of regr. =   0.32
Degrees of freedom for numer. = 2 , for denomin. = 22

F-value from table (alpha = 0.05) = 3.440000
Do you want a list of the actual values and forecasts ? Y

Continued

FIGURE F7.2   Some output from REGRE to predict college grade-point average as a function of SAT mathematics score and interview rating (cont.)

	Actual	Forecast	Residuals	% Error
1	2.8399999	3.176227	−0.33622748	−11.838996
2	3.0500000	2.9782972	0.71702659E-01	2.3509068
3	2.8599999	2.9947832	−0.13478333	−4.7127037
4	2.4000001	2.8488021	−0.44880196	−18.700081
5	3.5500000	3.3699560	0.18004394	5.0716600
6	3.0899999	3.3458505	−0.25585049	−8.2799520
7	3.7800000	3.1389823	0.64101768	16.958139
8	2.5799999	3.2423584	−0.66235858	−25.672813
9	2.4200001	2.4960105	−0.76010540E-01	−3.1409311
10	3.2800000	3.2711866	0.88133216E-02	0.26869884
11	3.1400001	2.9638252	0.17617482	5.6106629
12	3.1900001	2.8444846	0.34551540	10.831203
13	3.4100001	3.2981172	0.11188298	3.2810259
14	3.3499999	3.1991742	0.15082586	4.5022645
15	2.9100001	2.9684901	−0.58489978E-01	−2.0099649
16	2.5200000	2.8061969	−0.28619704	−11.357025
17	2.7000000	2.7568412	−0.56841224E-01	−2.1052306
18	2.6900001	2.9212782	−0.23127815	−8.5977001
19	3.4700000	3.5016968	−0.31696677E-01	−0.91344893
20	3.4600000	2.9777179	0.48228216	13.938790
21	3.1099000	3.1799791	−0.70079088E-01	−2.2534194
22	2.8199999	2.9062266	−0.86226761E-01	−3.0576866
23	3.0799999	3.3008261	−0.22082609	−7.1696782
24	3.5200000	3.1909311	0.32906884	9.3485470
25	3.4300001	2.9716628	0.45883374	13.362607

Durbin-Watson test calculated =   2.23
Durbin-Watson test from table (for 2 indep. variables and 25 obs.):
Upper = 1.55 Region A (Accept) : 1.55 to 2.45
Lower = 1.21 Region B (?)   : 1.21 to 1.55 and 2.45 to 2.79
                Region C (Reject) : 0   to 1.21 and 2.79 to 4.00

If we used this to predict the college performance of a student with an SAT mathematics score of 680 and an interview rating of 7.0, the predicted college grade-point average would be

$$\hat{y} = 0.709 + 0.003(680) + 0.115(7.0) = 3.55 \quad .$$

A student with an SAT mathematics score of 510 and an interview rating of 4.5, on the other hand, would have a predicted college grade-point average of

$$\hat{y} = 0.709 + 0.003(510) + 0.115(4.5) = 2.76 \quad .$$

The choice of one regression model over another is not always a clear-cut decision. In this example, the small t-statistics for tests concerning β_1 and β_2 caused us to drop x_1 from the model and try a new regression analysis with x_2 and x_3 as the only independent variables. Of course, the sample size of $n = 25$ is quite small. In practice, a college admissions officer has much more data to work with in attempting to come up with a good model for predicting college performance.

EXERCISES

F7.1.

Table F7.2 gives the number of accidents per year (in hundreds), the number of licensed vehicles (in thousands), and the size of the police force for 10 small communities. Use REGRE to conduct a multiple regression analysis that expresses the number of accidents as a function of the other two variables, and interpret the results.

TABLE F7.2   Number of accidents per year (in hundreds), number of licensed vehicles (in thousands), and size of police force for 10 communities

Accidents	Vehicles	Police
1	4	20
4	10	6
5	15	2
4	12	8
3	8	9
4	16	8
2	5	12
1	7	15
4	9	10
2	10	10

F7.2.

In Exercise F7.1, run a regression analysis with the number of accidents as the dependent variable and the number of licensed vehicles as the independent variable, and then run another regression analysis with the number of accidents as the dependent variable and the size of the police force as the independent variable. Compare these two regressions with the multiple regression from Exercise F7.1, and discuss.

F7.3.

Using the data in Table F7.3, run a multiple regression analysis with sales as the dependent variable and price and advertising as the independent variables. If the price

TABLE F7.3 Sales (in units of $10,000), price (in dollars), and advertising (in units of $1000) for a product

Sales	Price	Advertising
12	4	3
8	4	0
9	5	5
14	5	7
6	6	3
11	6	8
10	7	6
8	7	8

and advertising were 5 and 4, respectively, what would the predicted sales from the estimated multiple regression equation be?

Summary of Part F

Information about relationships among variables can be very helpful when we want to make predictions. In turn, the resulting predictions may play an important role in decision making. Of course, relationships can also be of interest simply because we would like to understand better whether certain variables are related to each other and, if so, how they are related. The purpose of Part F has been to study the statistical analysis of relationships via correlation and regression. Because of the importance of relationships among variables, correlation and regression are among the most widely used statistical techniques.

Linear functions are used to represent many relationships. Such functions are relatively easy to work with and provide reasonable approximations in many instances. Moreover, transformations of variables make it possible to consider nonlinear relationships within the framework of our linear statistical model. Correlations between variables (which can be transformed variables) tell us something about the direction and strength of relationships. Regression analysis provides a regression equation to represent the relationship between one variable and a set of one or more other variables. The assumptions underlying a regression model can be investigated, inferences can be made about the parameters of the regression model (the regression coefficients, the error variance) as well as the form of the model, and the regression equation can be used to generate predictions. More than any other part of this book, Part F provides some of the spirit of model-building that pervades good statistical practice.

VOCABULARY LIST

adjusted R^2

analysis of residuals

assumptions of linear regression

coefficient of determination

correlation

correlation coefficient

covariance

dependent variable

Durbin-Watson test

error term

estimated regression line

explanatory variable

exponential function

forecast

function
functional relationship
independent variable
intercept of regression line
least squares
linear function
linear regression
logarithmic function
multiple regression
negative relationship
nonlinear relationship
pairs of observations
population regression line
positive relationship

prediction
quadratic function
regression
regression coefficients
regression line
residual
scatter diagram
simple linear regression
slope of regression line
standard deviation of prediction error
standard deviation of the regression
standard error of regression coefficient
strength of relationship
transformation

LIST OF SYMBOLS

(x_i, y_i)	— a pair representing the values of x and y for the ith observation
cov (x,y)	— the covariance of x and y
r	— the sample correlation coefficient
ρ	— the population correlation coefficient
b_0	— the y-intercept of the estimated regression line
b_1	— the slope of the estimated regression line
e_i	— the error in using a regression line to predict y_i
$\hat{y}$	— the predicted value of y from a regression line
β_0	— the y-intercept of the population regression line
β_1	— the slope of the population rgression line
σ_e^2	— the error variance
s_e^2	— the estimated error variance
s_e	— the standard deviation of the regression
R^2	— the coefficient of determination
D	— the Durbin-Watson statistic
d_L, d_U	— critical values of the Durbin-Watson statistic
k	— the number of independent variables in regression

x_i – the ith observation of variable x, or the ith independent variable in multiple regression

x_{ij} – the jth observation of variable x_i

$(y_i, x_{1i}, x_{2i}, \ldots, x_{ki})$ – the ith observation of y and k independent variables $x_1, \ldots, x_k$

$b_0, b_1, b_2, \ldots, b_k$ – the estimated regression coefficients in a multiple regression

$\beta_0, \beta_1, \beta_2, \ldots \beta_k$ – the population regression coefficients in a multiple regression

FORMULAS

$$\mathrm{cov}(x,y) = \frac{\displaystyle\sum_{i=1}^{n} (x_i - \bar{x})(y_i - \bar{y})}{n - 1}$$

$$r = \frac{\mathrm{cov}(x,y)}{s_x s_y}$$

$$r = \frac{n\Sigma x_i y_i - (\Sigma x_i)(\Sigma y_i)}{\sqrt{[n\Sigma x_i^2 - (\Sigma x_i)^2]\,[n\Sigma y_i^2 - (\Sigma y_i)^2]}}$$

$$-1 \leqslant r \leqslant 1$$

$$t = \frac{r\sqrt{n-2}}{\sqrt{1-r^2}} \qquad\qquad \text{(for test of } H_0\text{: } \rho = 0)$$

$$y = b_0 + b_1 x$$

$$y_i = b_0 + b_1 x_i + e_i$$

$$\hat{y}_i = b_0 + b_1 x_i$$

$$e_i = y_i - \hat{y}_i$$

$$b_1 = \frac{n\Sigma x_i y_i - (\Sigma x_i)(\Sigma y_i)}{n\Sigma x_i^2 - (\Sigma x_i)^2}$$

$$b_0 = \frac{\Sigma y_i - b_1 \Sigma x_i}{n}$$

$$b_1 = r\frac{s_y}{s_x}$$

$$y = \beta_0 + \beta_1 x$$

$$s_e^2 = \frac{\sum_{i=1}^{n} e_i^2}{n - 2}$$

$$s_e = \sqrt{\frac{\sum_{i=1}^{n} e_i^2}{n - 2}}$$

$$t = \frac{b_0 - \beta_0}{s_e \left(\sqrt{\sum x_i^2/n}\right)/s_x\sqrt{n - 1}} \qquad \text{(for inferences concerning } \beta_0\text{)}$$

$$t = \frac{b_1 - \beta_1}{s_e/s_x \sqrt{n - 1}} \qquad \text{(for inferences concerning } \beta_1\text{)}$$

$$E = t_{\alpha/2} \frac{s_e \sqrt{\sum x_i^2/n}}{s_x \sqrt{n - 1}} \qquad \text{(for estimation of } \beta_0\text{)}$$

$$E = t_{\alpha/2} \frac{s_e}{s_x \sqrt{n - 1}} \qquad \text{(for estimation of } \beta_1\text{)}$$

$$b_0 \pm t_{\alpha/2} \frac{s_e \sqrt{\sum x_i^2/n}}{s_x \sqrt{n - 1}} \qquad \text{(confidence interval for } \beta_0\text{)}$$

$$b_1 + t_{\alpha/2} \frac{s_e}{s_x \sqrt{n - 1}} \qquad \text{(confidence interval for } \beta_1\text{)}$$

$$F = \frac{R^2}{(1 - R^2)/(n - 2)} \qquad \text{(for goodness of fit in simple regression)}$$

$$s_{e_i} = s_e \sqrt{1 + \frac{1}{n} + \frac{(x_i - \bar{x})^2}{s_x^2 (n - 1)}}$$

$$t = \frac{\hat{y}_i - y_i}{s_{e_i}} \qquad \text{(for prediction of } y_i\text{)}$$

$$\hat{y}_i \pm t_{\alpha/2} s_{e_i} \qquad \text{(interval for } y_i\text{)}$$

$$D = \frac{\sum_{i=2}^{n} (e_i - e_{i-1})^2}{\sum_{i=1}^{n} e_i^2}$$

$$y = b_0 + b_1 x_1 + b_2 x_2 + \ldots + b_k x_k$$

$$\hat{y}_i = b_0 + b_1 x_{1i} + b_2 x_{2i} + \ldots + b_k x_{ki}$$

$$y = \beta_0 + \beta_1 x_1 + \beta_2 x_2 + \ldots + \beta_k x_k$$

$$F = \frac{R^2/k}{(1 - R^2)/(n - k - 1)} \qquad \text{(for goodness of fit in multiple regression)}$$

RELEVANT ISP COMMANDS

The two most relevant commands for Part F are SCVCR and REGRE. These commands perform all of the necessary computations for correlation and regression, even providing for scatter diagrams (in SCVCR) and checks for the applicability of the regression assumptions (in REGRE).

Graphs can be very handy in correlation and regression. Scatter diagrams can be obtained directly by using the command PSCAT. Various plots useful in checking regression assumptions can be generated by PLOT. Also, the normality of the error terms can be tested with the goodness of fit test given by PDIST.

As in the preceding parts of this book, the C commands are useful for any computations. As for the T commands, transformations are of particular interest in regression because they enable us to consider nonlinear relationships. Thus, the transformations available in ISP are especially useful in regression.

The MR commands in ISP provide brief summaries of the concepts of regression and correlation. The seven MR modules correspond to the seven chapters of Part F.

MRF1:   Functional relationships
MRF2:   Covariance and correlation
MRF3:   Linear regression
MRF4:   Inferences in linear regression
MRF5:   Testing the assumptions of regression
MRF6:   Transformations and nonlinear relationships
MRF7:   Multiple regression

PART G

Forecasting

An important role of statistical techniques is to help us make predictions, or forecasts, about what is likely to happen in the future. For example, we might want a forecast for interest rates on home mortgages at the end of the year, for sales of a particular product next month, for the number of traffic deaths under different speed limits, or for the average score of a school's graduating class on an achievement test. Forecasts can be extremely useful for decision making and planning.

In Part F, we discussed the use of regression not just to represent relationships among variables, but also to make forecasts. In Part G, we will consider other methods that are valuable in forecasting. By analyzing historical observations of variables over time, we can attempt to discover patterns. Information about such patterns can be helpful when we want to forecast what might happen in the future. The choice of a forecasting method in a particular situation is influenced by the types of patterns that are found in the past data and the degree to which we expect these patterns to persist or change in the future.

Chapter G1 provides an introduction to time series data, which is what we focus upon in Part G. Chapter G2 describes summary measures, autocorrelations and cross-autocorrelations, that provide important information about patterns in time series data. The decomposition of a time series into components representing trends, cycles, seasonality, and randomness is discussed in Chapter G3. The final chapter in Part G, Chapter G4, involves a commonly-used forecasting technique called exponential smoothing.

G1

Time Series

Often data are collected at a particular point in time. For instance, if an examination is given to a class of 100 students, the 100 scores represent 100 data points, or observations. Statistical methods discussed in previous parts of this book enable us to analyze such data. In Part G we are concerned with time series data: observations that are made over time. With time series data, successive data points represent observations adjacent in time (adjacent days, weeks, months, quarters, years, or some other time periods). Therefore, we might expect successive observations in time series data to be related in some manner. In contrast, successive observations in the exam-scores illustration are scores of two students, and there is no reason to suspect a particular relationship between these scores. Relationships among observations in time series data can be investigated and exploited to forecast future values in the series.

A listing of data from a time series in order of observation can provide some idea of systematic patterns that exist in the series. A graph in which the observations are plotted in order as a function of time is even more useful than a listing. The graph makes it easier to spot patterns and to get an impression of the characteristics of the time series. Similar graphs of differences between successive observations (first differences) or differences between successive first differences (second differences) can also be helpful. If we let $y_1, y_2, \ldots, y_n$ denote the values in the time series, with y_1 representing the earliest observation and y_n representing the last observation, then the first difference at time i (when y_i is observed) is

$$d_i = y_i - y_{i-1} \qquad\qquad \text{(G1–1)}$$

and the second difference at time i is $d_i - d_{i-1}$. Summary measures that supplement the graphical analysis and tell us more about the characteristics of a time series will be discussed in the following chapters.

ISP COMMANDS

The command PLOT plots up to three variables on a time scale, enabling you to observe patterns in individual series and to see how different series might be related to each other. Also, a transformation, TDIF, is available in ISP to provide the first or second differences in a series of data. Commands such as PFREQ and PHIST can be used to provide further information about the series. Commands providing summary measures of particular interest for time series data will be covered in the next two chapters.

SOLVED EXAMPLE

Forecasts of sales of a product are needed to enable manufacturers to decide how many units to produce and retailers to decide how many units to keep in inventory. The data in Table G1.1 represent quarterly sales of a particular brand of paper (in thousands of units) for 51 successive quarters (three-month periods). For example, the observation 3017.60 in the first quarter indicates sales of 3,017,600 units in that quarter.

The plot of quarterly sales against time in Figure G1.1 is much easier to interpret than the listing of the data in Table G1.1. We can see that there is a tendency for sales to increase somewhat over time. For instance, for the first 8 quarters (2 years) of data, the sales figures are in the 2000–3500 range; for the last 8 quarters, they are in the 3500–6500 range. In forecasting terminology, a tendency for the values in a series to increase over time is called a positive trend.

TABLE G1.1   Quarterly sales of paper (in thousands of units) for 51 quarters

Quarter	Sales	Quarter	Sales	Quarter	Sales
1	3017.60	18	3701.19	35	3172.18
2	3043.54	19	2642.38	36	4223.76
3	2094.35	20	3585.52	37	4690.48
4	2809.84	21	4078.66	38	4694.48
5	3274.80	22	3907.06	39	3342.35
6	3163.28	23	2818.46	40	4577.63
7	2114.30	24	4089.50	41	4965.46
8	3024.57	25	4339.61	42	5026.05
9	3327.48	26	4148.60	43	3470.14
10	3493.49	27	2916.45	44	4525.94
11	2439.93	28	4084.64	45	5258.71
12	3490.79	29	4242.42	46	5189.58
13	3685.08	30	3997.58	47	3596.76
14	3661.23	31	2881.01	48	3881.60
15	2378.43	32	4036.23	49	6231.20
16	3459.55	33	4360.33	50	6325.40
17	3849.63	34	4360.53	51	3602.20

FIGURE G1.1   Plot of quarterly sales of paper (in thousands of units) for 51 quarters

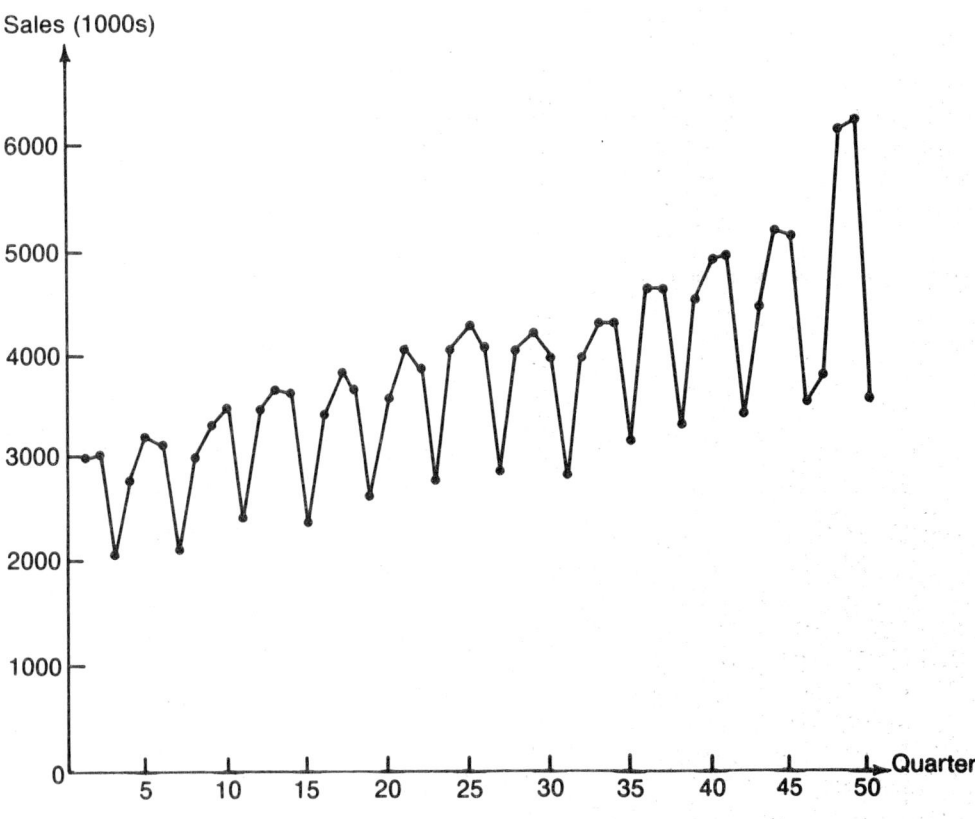

Another pattern that is clear from Figure G1.1 is a distinct seasonality in the series. The series drops down to low values in the third, seventh, eleventh, and so on, quarters. These low values occur in the third quarter of each year, indicating that this quarter is not a good period in terms of sales of paper.

To learn more about the series, we can compute the first differences. For instance, the first differences in quarters 2 and 3 are, from (G4–1),

$$d_2 = y_2 - y_1 = 3043.54 - 3017.60 = 25.94$$
$$\text{and } d_3 = y_3 - y_2 = 2094.35 - 3043.54 = -949.19 \quad .$$

The complete set of first differences for the sales series is given in Table G1.2 and graphed in Figure G1.2. From the graph, we can see that taking the first differences has removed the trend from the series. The plot of first differences varies about zero, with no apparent

TABLE G1.2 Computation of first differences in series of quarterly sales of paper

Quarter (i)	y_i	y_{i-1}	Difference
1	3017.60	—	—
2	3043.54	3017.60	25.94
3	2094.35	3043.54	−949.19
4	2809.84	2094.35	715.49
5	3274.80	2809.84	464.96
6	3163.28	3274.80	−111.52
7	2114.30	3163.28	−1048.98
8	3024.57	2114.30	910.27
9	3327.48	3024.57	302.91
10	3493.49	3327.48	166.01
11	2439.93	3493.49	−1053.56
12	3490.79	2439.93	1050.86
13	3685.08	3490.79	194.29
14	3661.23	3685.08	−23.85
15	2378.43	3661.23	−1282.80
16	3459.55	2378.43	1081.12
17	3849.63	3459.55	390.08
18	3701.19	3849.63	−148.44
19	2642.38	3701.19	−1058.81
20	3585.52	2642.38	943.14
21	4078.66	3585.52	493.14
22	3907.06	4078.66	−171.60
23	2818.46	3907.06	−1088.60
24	4089.50	2818.46	1271.04
25	4339.61	4089.50	250.11
26	4148.60	4339.61	−191.01
27	2916.45	4148.60	−1232.15
28	4084.64	2916.45	1168.19
29	4242.42	4084.64	157.78
30	3997.58	4242.42	−244.84
31	2881.01	3997.58	−1116.57
32	4036.23	2881.01	1155.22
33	4360.33	4036.23	324.10
34	4360.53	4360.33	0.20
35	3172.18	4360.53	−1188.35
36	4223.76	3172.18	1051.58
37	4690.48	4223.76	466.72
38	4694.48	4690.48	4.00
39	3342.35	4694.48	−1352.13
40	4577.63	3342.35	1235.28
41	4965.46	4577.63	387.83
42	5026.05	4965.46	60.59
43	3470.14	5026.05	−1555.91
44	4525.94	3470.14	1055.80
45	5258.71	4525.94	732.77
46	5189.58	5258.71	−69.13

Continued

TABLE G1.2   Computation of first differences in series of quarterly sales of paper (cont.)

47	3596.76	5189.58	−1592.82
48	3881.60	3596.76	284.84
49	6231.20	3881.60	2349.60
50	6325.40	6231.20	94.20
51	3602.20	6325.40	−2723.20

tendency to increase or decrease over time. The seasonality, however, is still present, with the low sales figures in the third quarter of each year translating into substantial negative first differences in each third quarter.

FIGURE G1.2   Plot of first differences for series of quarterly sales of paper

First Difference (1000s)

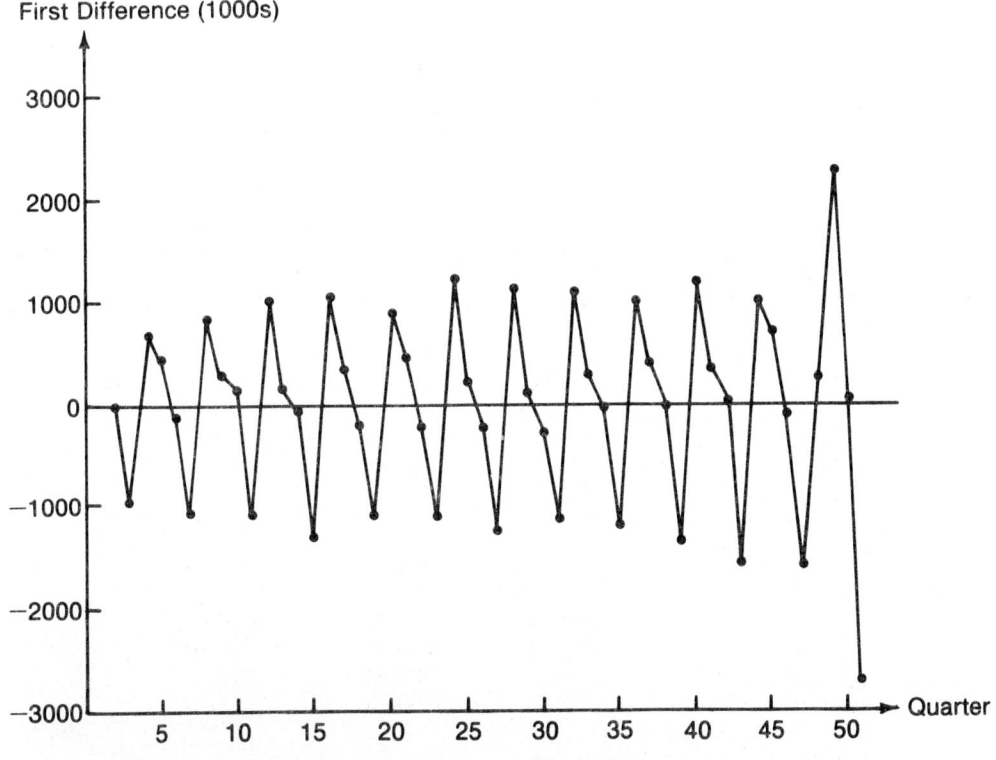

EXERCISES

G1.1.

The high temperatures in degrees Fahrenheit in Durham, North Carolina on the first 20 days of May one year are given in Table G1.3. Use PLOT to plot this time series of temperatures. Also, find the first differences by using TDIF and then plot the first differences.

TABLE G1.3 **High temperatures (degrees Fahrenheit) in Durham, North Carolina on the first 20 days of May**

Day	Temp.	Day	Temp.	Day	Temp.	Day	Temp.
1	64	6	75	11	72	16	79
2	68	7	67	12	75	17	77
3	65	8	66	13	79	18	74
4	73	9	69	14	81	19	75
5	73	10	68	15	78	20	78

G1.2.

The yearly inflation rate and unemployment rate, both expressed in percentages, for a particular region over a 20-year period are provided in Table G1.4. Enter both series in ISP and use PLOT to plot both series on the same graph. Find the first differences for the two series and plot the two sets of first differences.

TABLE G1.4 **Inflation rate and unemployment rate in percentages for a region over a 20-year period**

Year	Inflat.	Unempl.	Year	Inflat.	Unempl.	Year	Inflat.	Unempl.
1	4	4	8	10	5	15	16	11
2	3	5	9	11	6	16	14	11
3	5	4	10	13	7	17	12	10
4	5	4	11	14	7	18	11	8
5	8	5	12	14	9	19	11	7
6	7	4	13	12	10	20	10	7
7	7	6	14	15	10			

G1.3.

Table G1.5 presents the monthly road casualties in a country for a five-year period covering the 60 months from January 1980 to December 1984. Plot this series and the series of first differences.

TABLE G1.5   Monthly road casualties in a country for the period from January 1980 to December 1984

Month	Casualties	Month	Casualties	Month	Casualties
1	260	21	344	41	341
2	245	22	339	42	309
3	279	23	339	43	347
4	291	24	364	44	337
5	347	25	270	45	336
6	331	26	263	46	310
7	360	27	298	47	289
8	375	28	326	48	297
9	348	29	351	49	239
10	355	30	344	50	247
11	334	31	357	51	275
12	329	32	336	52	267
13	290	33	319	53	287
14	247	34	351	54	303
15	313	35	334	55	313
16	324	36	376	56	321
17	339	37	275	57	312
18	350	38	272	58	314
19	364	39	302	59	308
20	365	40	286	60	306

G2

Autocorrelation

The correlation coefficient, introduced and used extensively in Part F, provides useful information about the relationship between two variables. In the analysis of time series data, a particular type of correlation coefficient, called an autocorrelation coefficient, helps us understand how the data behave over time. If we have n observations y_1, y_2, ..., y_n over time, where y_1 represents the earliest observation and y_n the most recent observation, then the autocorrelation coefficient of lag k is

$$r_k = \frac{\sum_{i=1}^{n-k}(y_i - \bar{y})(y_{i+k} - \bar{y})}{\sum_{i=1}^{n}(y_i - \bar{y})^2} . \qquad (G2\text{--}1)$$

The expression "of lag k" indicates that we are considering the relationship between a variable (y) and the same variable k time periods earlier. Thus, r_k tells us how values of the same variable that are k time periods apart relate to each other. The closer r_k is to zero, the weaker the relationship; values of r_k near 1 or -1 indicate strong positive and negative relationships, respectively.

For a given series of data, autocorrelation coefficients can be computed for different lags k. A graph of r_k as a function of k often reveals important characteristics of the series. Such a graph is called an autocorrelation function. Furthermore, in addition to the autocorrelation function for the original data, we can compute autocorrelations and graph autocorrelation functions for first differences or higher-order differences in the data.

Often we have time series data for two or more variables over the same time periods. A measure of the relationship between one variable and a lagged second variable is called a cross-autocorrelation coefficient. The cross-autocorrelation coefficient of y with a second variable x that is lagged k periods is

$$r_{xy}(k) = \frac{\sum_{i=1}^{n-k}(x_i - \bar{x})(y_{i+k} - \bar{y})}{\sqrt{\left[\sum_{i=1}^{n}(x_i - \bar{x})^2\right]\left[\sum_{i=1}^{n}(y_i - \bar{y})^2\right]}} \ . \qquad (G2\text{--}2)$$

When k=0, this is simply the ordinary correlation between x and y. When k is positive, y is being correlated with a lagged version of x. A negative k indicates that x is being correlated with a lagged version of y. Thus, a plot of $r_{xy}(k)$ as a function of k, including both positive and negative values of k, provides a considerable amount of information about relationships between the series of x-values and the series of y-values.

ISP COMMANDS

The command FAUTO provides autocorrelation coefficients for any variable of your choice and also gives a plot of the autocorrelation function. The maximum value of k can be specified by the user or the choice can be left to the computer. Autocorrelations and autocorrelation functions for the first and second differences in a series of data can also be obtained from FAUTO if desired.

Another command, FCROS, can be used to obtain cross-autocorrelation coefficients and a plot of these coefficients as a function of k, including negative k as well as positive k and k=0. The cross-autocorrelations can be computed for the original data, for seasonally adjusted data (to be discussed in Chapter G3), or for differenced and seasonally adjusted data.

SOLVED EXAMPLE

The data in Table G1.1 represent a time series of quarterly sales of paper for 51 quarters. A look at the series and the first differences indicates the presence of some systematic patterns, as discussed in Chapter G1. The measures introduced here in Chapter G2 might shed further light on these patterns.

The autocorrelation function generated by FAUTO for lags of 1 through 15 is shown in Figure G2.1. The seasonal effect can be seen clearly in terms of the high values of r_4, r_8, and r_{12}. The highest autocorrelation is $r_4 = 0.734$, indicating that an observation is most closely related with sales in the same quarter a year earlier or later. The next highest autocorrelation is $r_8 = 0.567$, demonstrating that the relationship of an observation with sales in the same season *two* years earlier or later is still reasonably strong. The relationship of sales with the previous period's sales is moderate ($r_1 = 0.450$), but

FIGURE G2.1   Autocorrelation function from FAUTO for series of quarterly sales of paper

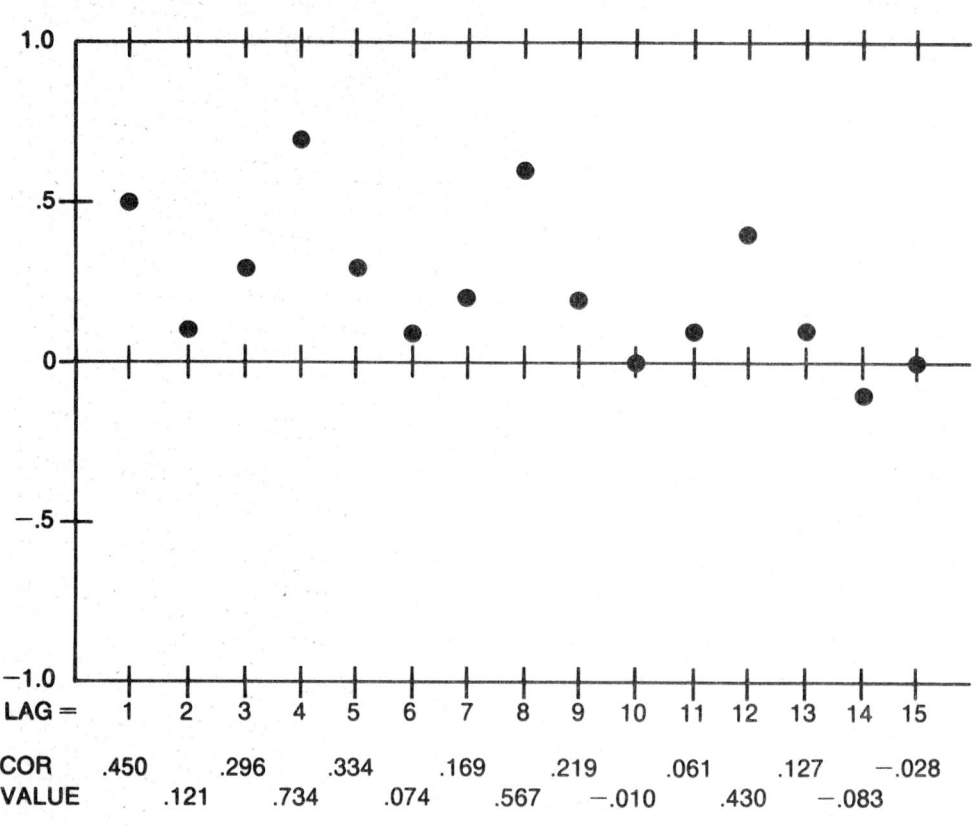

LAG =	1	2	3	4	5	6	7	8	9	10	11	12	13	14	15
COR	.450		.296		.334		.169		.219		.061		.127		−.028
VALUE		.121		.734		.074		.567		−.010		.430		−.083	

most of the other autocorrelations, although positive, are relatively small. The dominant feature of the autocorrelation function is the seasonal effect.

The command FAUTO can also be used to find the autocorrelation function for first differences in the sales series, and this autocorrelation function is presented in Figure G2.2. As in the case of Figure G2.1, the most striking feature of Figure G2.2 is the seasonal effect. The autocorrelations r_4, r_8, and r_{12} are all positive and reasonably high (0.612, 0.542, and 0.486, respectively). All other autocorrelations are negative and relatively small.

To investigate possible relationships between the price of paper and sales, FCROS can be used to find cross-autocorrelations. The plot of cross-autocorrelations given in Figure G2.3 reveals that the ordinary correlation $r_{xy}(0)$ with no lags is high (0.765). The

FIGURE G2.2   Autocorrelation function from FAUTO for first differences in series of quarterly sales of paper

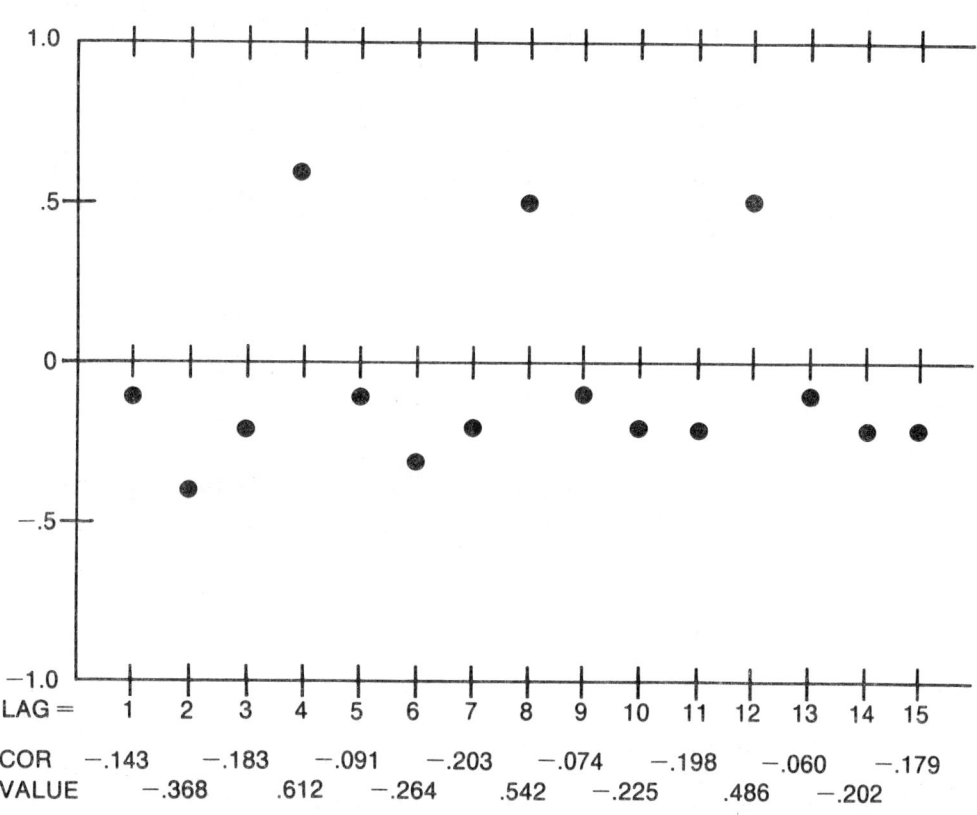

LAG =	1	2	3	4	5	6	7	8	9	10	11	12	13	14	15
COR	−.143		−.183		−.091		−.203		−.074		−.198		−.060		−.179
VALUE		−.368		.612		−.264		.542		−.225		.486		−.202	

price would tend to increase over time due to inflation, and the sales growth for this brand of paper is also relatively consistent, as the positive trend in Figure G1.1 indicates. Thus, the high value of $r_{xy}(0)$ can be explained in terms of a time effect and does not imply that in a given period, sales would be positively related to price. The cross-autocorrelations, values of $r_{xy}(k)$ for k other than zero, are all close to zero in Figure G2.3. It is doubtful that relationships between sales and lagged price or price and lagged sales are strong enough to be of much interest.

FIGURE G2.3   Cross-autocorrelations of price of paper and quarterly sales of paper from FCROS

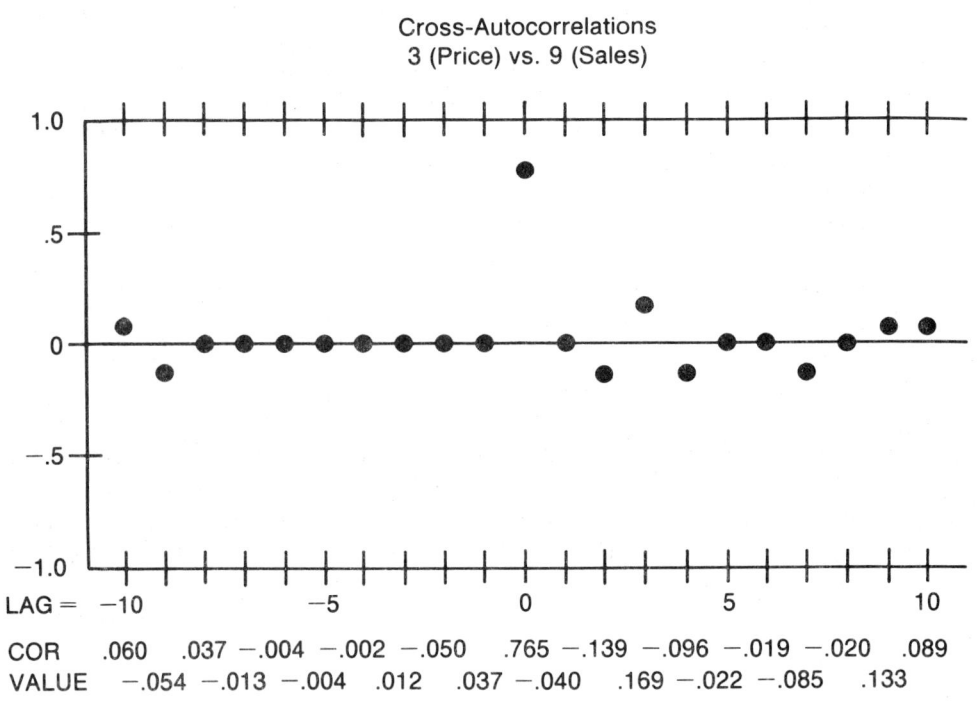

Cross-Autocorrelations
3 (Price) vs. 9 (Sales)

COR	.060	.037	−.004	−.002	−.050	.765	−.139	−.096	−.019	−.020	.089
VALUE		−.054	−.013	−.004	.012	.037	−.040	.169	−.022	−.085	.133

EXERCISES

G2.1.
Use FAUTO to find the autocorrelation function for the series of temperatures given in Table G1.3.

G2.2.
Generate the autocorrelation function for the series of monthly road casualties given in Table G1.5. What does the autocorrelation function tell you about this series?

G2.3.
For the data in Table G1.4, use FAUTO to find the autocorrelation functions for the inflation series and the unemployment series. Also, use FCROS to find the cross-auto-correlations and attempt to interpret these results.

G3

Decomposition of a Time Series

An individual observation that is part of a time series may be influenced by a number of components. In analyzing a time series, therefore, we often find it helpful to attempt to decompose the series into certain components. The four components usually considered are trend, cycle, seasonality, and randomness. Typically a multiplicative model is used, with an observation y_i being expressed as a product of the terms T_i (trend), C_i (cycle), S_i (seasonality), and R_i (randomness):

$$y_i = T_i C_i S_i R_i \quad . \tag{G3-1}$$

The trend component T_i corresponding to the ith observation is expressed in the same units as y_i. The other three components are expressed in relative terms. For example, values of C_i or S_i above one indicate positive cyclical or seasonal effects that tend to make y_i greater than T_i, and values below one suggest that y_i is likely to be less than the trend component. The randomness component simply represents all factors other than trend, cycle, and seasonality influencing the values in the time series.

An approach known as classical decomposition is often used to decompose a series. First, an attempt is made to isolate seasonal effects. The length of the seasonality (the number of ''seasons'') may be obvious from the nature of the data (12 for monthly data, 4 for quarterly data, and so on) or may be determined from the autocorrelation function. If the length of seasonality is denoted by L, we calculate a moving average with a ''window'' of L observations. The value of the moving average m_i corresponding to time period i is simply the average of the L observations centered at time period i. Averaging over a full set of seasons eliminates the seasonal effect and much of the randomness. Therefore, we view the series of moving averages as representing $T_i C_i$. Dividing the observations y_i by the corresponding moving averages m_i results in ratios reflecting seasonality and randomness. Averaging these ratios for each season (for example, averaging all of the ratios for January, all those for February, and so on with monthly data) eliminates much of the randomness and gives us an estimate of the seasonal effects for the L seasons.

Given the seasonal effects, we can deseasonalize the data by dividing each observation y_i by the corresponding seasonal component S_i. The combined trend-cycle effect, sometimes referred to as just the trend effect if cyclical patterns are not treated separately, is estimated by a regression with the time period i as the independent variable and the deseasonalized data as the dependent variable. Regression models linear or nonlinear in time can be used. Forecasts of future values of the time series can then be determined by using the regression (or perhaps some other extrapolation technique) to forecast future deseasonalized values and then multiplying these forecasts by the appropriate seasonal factors to generate the final forecasts.

ISP COMMANDS

The command FDEC applies the classical decomposition approach to a time series. Moving averges are provided, estimates of seasonal effects are calculated, and some accuracy measures indicating how closely the model fits the time series are given. In addition, FDEC can be used to generate forecasts for future values of the series with a number of options for extrapolating the trend-cycle effect before multiplying by the seasonal factors.

SOLVED EXAMPLE

Some output from the use of FDEC to apply the classical decomposition technique to the series of quarterly sales of paper (the data from Table G1.1) is presented in Figure G3.1. Since the length of seasonality for this quarterly data is L = 4 periods, moving averages of four observations are computed. For instance, the first two moving averages for this series are

$$\frac{3017.60 + 3043.54 + 2094.35 + 2809.84}{4} = 2741.33$$

and
$$\frac{3043.54 + 2094.35 + 2809.84 + 3274.80}{4} = 2805.63 \quad .$$

Because L is even, no moving average is centered exactly on a single period; the first moving average is centered midway between the second and third quarters, for example. To remedy this situation, we compute a centered moving average by averaging the two

FIGURE G3.1   Some output from FDEC for decomposition of series of quarterly sales of paper

***** Classical Decomposition *****

What is the length of seasonality ? 4
Do you want :
  1. All output
  2. Trend-Cycles only
  3. Seasonal Indices only. (answer 1, 2, or 3) ? 1

Period	Original data	Moving average	Centered mov.avg.	Ratios
1	3017.60	—	—	—
2	3043.54	—	—	—
3	2094.35	2741.33	2773.48	0.755
4	2809.84	2805.63	2820.60	0.996
5	3274.80	2835.57	2838.06	1.154
6	3163.28	2840.55	2867.40	1.103
7	2114.30	2894.24	2900.82	0.729
8	3024.57	2907.41	2948.68	1.026
9	3327.48	2989.96	3030.66	1.098
10	3493.49	3071.37	3129.65	1.116
11	2439.93	3187.92	3232.62	0.755
12	3490.79	3277.32	3298.29	1.058
13	3685.08	3319.26	3311.57	1.113
14	3661.23	3303.88	3299.98	1.109
15	2378.43	3296.07	3316.64	0.717
16	3459.55	3337.21	3342.21	1.035
17	3849.63	3347.20	3380.19	1.139
18	3701.19	3413.19	3428.93	1.079
19	2642.38	3444.68	3473.31	0.761
20	3585.52	3501.94	3527.67	1.016
21	4078.66	3553.41	3575.42	1.141
22	3907.06	3597.43	3660.42	1.067
23	2818.46	3723.42	3756.04	0.750
24	4089.50	3788.66	3818.85	1.071
25	4339.61	3849.04	3861.29	1.124
26	4148.60	3873.54	3872.93	1.071
27	2916.45	3872.33	3860.18	0.756
28	4084.64	3848.03	3829.15	1.067
29	4242.42	3810.27	3805.84	1.115
30	3997.58	3801.41	3795.36	1.053
31	2881.01	3789.31	3804.05	0.757
32	4036.23	3818.79	3864.16	1.045
33	4360.33	3909.53	3945.92	1.105
34	4360.53	3982.32	4005.76	1.089

Continued

FIGURE G3.1   Some output from FDEC for decomposition of series of quarterly sales of paper (cont.)

35	3172.18	4029.20	4070.47	0.779
36	4223.76	4111.74	4153.48	1.017
37	4690.48	4195.23	4216.50	1.112
38	4694.48	4237.77	4282.00	1.096
39	3342.35	4326.23	4360.61	0.766
40	4577.63	4394.98	4436.43	1.032
41	4965.46	4477.87	4493.85	1.105
42	5026.05	4509.82	4503.36	1.116
43	3470.14	4496.90	4533.55	0.765
44	4525.94	4570.21	4590.65	0.986
45	5258.71	4611.09	4626.92	1.137
46	5189.58	4642.75	4562.21	1.138
47	3596.76	4481.66	4603.22	0.781
48	3881.60	4724.79	4866.76	0.798
49	6231.20	5008.74	5009.42	1.244
50	6325.40	5010.10	—	—
51	3602.20	—	—	—

Smallest Value

1.10	1.05	0.72	0.80

Year

3	8	4	12

Largest Value

1.24	1.14	0.78	1.07

Year

13	12	12	6

Season	Medial Average	Seasonal Indices
1	112.44	112.33
2	109.42	109.32
3	75.74	75.67
4	102.78	102.68
	400.376	400.000

Average Percentage Changes:

Original	Deseasonal	Random	Trend-Cycle
20.86	4.56	4.01	1.55

A C C U R A C Y Measures of M O D E L Fitting

Mean percentage error, or bias	=	−0.39
Mean absolute percentage error	=	2.53
Root mean squared error	=	197.3
R-square	=	0.953

moving averages closest to a given period. For quarter 3, this centered moving average is

$$\frac{2741.33 + 2805.63}{2} = 2773.48 \quad .$$

The ratios in the right-hand column of Figure G3.1 are ratios of the original observations divided by the corresponding moving averages. For the third quarter, the ratio is

$$2094.35/2773.48 = 0.755 \quad .$$

For each season, the smallest and largest ratios are found and displayed in Figure G3.1. For example, among all first quarters, the smallest ratio is 1.10 in the third year and the largest ratio is 1.24 in the thirteenth year. These extreme values are deleted and the average of the remaining ratios is computed for each season and multiplied by 100. The results, called medial averages, do not quite add up to 400 because the extreme ratios were dropped. Therefore, each medial average is divided by the sum of all four medial averages to arrive at the four seasonal indices given in Figure G3.1. The low value of the seasonal index for the third quarter reflects the fact that third-quarter sales tend to be particularly low, as discussed in Chapter G1.

To deseasonalize, or remove the seasonal effect from, the time series, we divide each observation by the appropriate seasonal index. The first-quarter sales figure for each year is divided by 1.1233 (the first-quarter seasonal index divided by 100), the second-quarter value is divided by 1.0932, the third-quarter observation by 0.7567, and the fourth-quarter sales by 1.0268. The results of these calculations are given in Table G3.1. The right-hand column, headed y_i/S_i, gives the deseasonalized sales values. As you can see from the plot in Figure G3.2, the deseasonalized series does not have the systematic seasonal variation shown in Figure G1.1 for the original data. Other than the last few observations, which are more extreme than the earlier values, the deseasonalized series seems quite stable with a positive trend.

Forecasts of sales for the next six quarters (quarters 52 through 57) are given in the output from FDEC in Figure G3.3. An exponential extrapolation of the trend-cycle effect was chosen, and the program gives point forecasts and limits for a 95 percent forecast interval. We can see that the considerable seasonal effect is taken into account in the forecasts; the forecast for quarter 55, which is the third quarter of a year, is particularly low, just as the observations for previous third quarters were low.

TABLE G3.1   Division by seasonal component S_i to deseasonalize sales data

Quarter (i)	y_i	S_i	y_i/S_i
1	3017.60	1.1233	2686.37
2	3043.54	1.0932	2784.07
3	2094.35	0.7567	2767.74
4	2809.84	1.0268	2736.50
5	3274.80	1.1233	2915.34
6	3163.28	1.0932	2893.60
7	2114.30	0.7567	2794.11
8	3024.57	1.0268	2945.63
9	3327.48	1.1233	2962.24
10	3493.49	1.0932	3195.66
11	2439.93	0.7567	3224.44
12	3490.79	1.0268	3399.68
13	3685.08	1.1233	3280.58
14	3661.23	1.0932	3349.09
15	2378.43	0.7567	3143.16
16	3459.55	1.0268	3369.25
17	3849.63	1.1233	3427.07
18	3701.19	1.0932	3385.65
19	2642.38	0.7567	3491.98
20	3585.52	1.0268	3491.94
21	4078.66	1.1233	3630.96
22	3907.06	1.0932	3573.97
23	2818.46	0.7567	3724.67
24	4089.50	1.0268	3982.76
25	4339.61	1.1233	3863.27
26	4148.60	1.0932	3794.91
27	2916.45	0.7567	3854.17
28	4084.64	1.0268	3978.03
29	4242.42	1.1233	3776.75
30	3997.58	1.0932	3656.77
31	2881.01	0.7567	3807.33
32	4036.23	1.0268	3930.88
33	4360.33	1.1233	3881.71
34	4360.53	1.0932	3988.77
35	3172.18	0.7567	4192.12
36	4223.76	1.0268	4113.52
37	4690.48	1.1233	4175.63
38	4694.48	1.0932	4294.26
39	3342.35	0.7567	4417.01
40	4577.63	1.0268	4458.15
41	4965.46	1.1233	4420.42
42	5026.05	1.0932	4597.56
43	3470.14	0.7567	4585.89
44	4525.94	1.0268	4407.81
45	5258.71	1.1233	4681.48
46	5189.58	1.0932	4747.15

Continued

TABLE G3.1   Division by seasonal component S_i to deseasonalize sales data (cont.)

47	3596.76	0.7567	4753.22
48	3881.60	1.0268	3780.29
49	6231.20	1.1233	5547.23
50	6325.40	1.0932	5786.13
51	3602.20	0.7567	4760.41

FIGURE G3.2   Plot of deseasonalized quarterly sales of paper

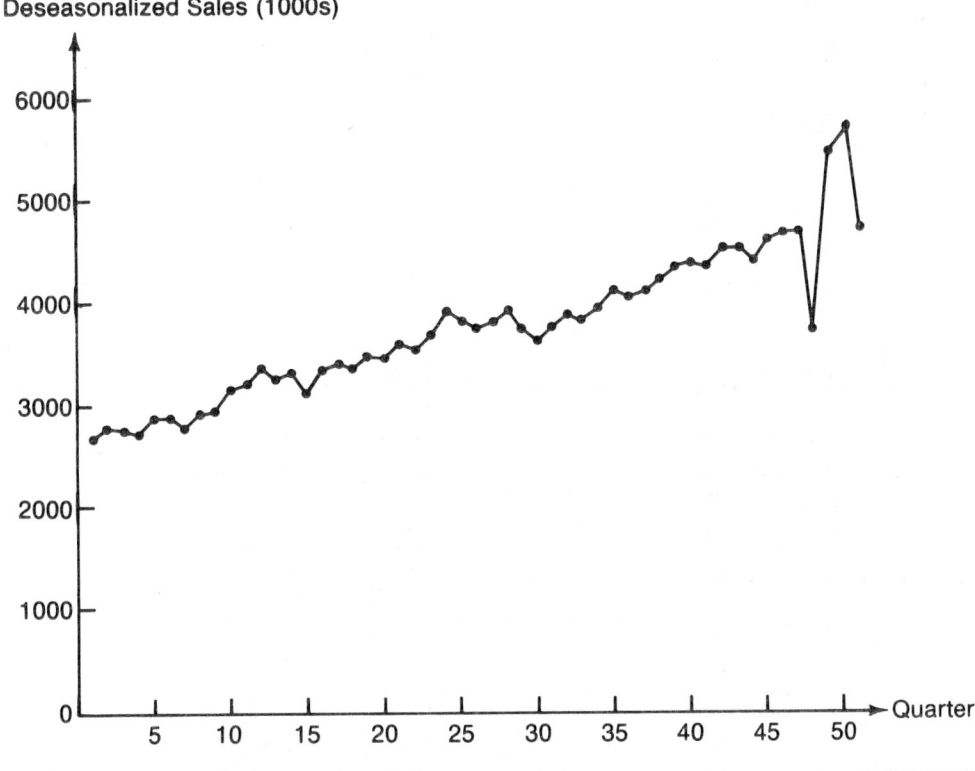

FIGURE G3.3   Some output from FDEC with forecasts of quarterly sales of paper

Number of forecasts ? 6
Do you want to input your own estimates of the Trend-Cycle ? : N
   Do you want estimates of the Trend-Cycle by :
     1. Linear extrapolation
     2. Exponential extrapolation
     3. Average growth increase
     4. Average growth increase of last two years
     5. A growth rate specified by yourself   (answer 1, 2, 3, 4, or 5) ? 2

Period	Low Limit	F O R E C A S T	Upper Limit
52	4722.93	5083.68	5444.43
53	5051.35	5624.84	6198.34
54	4549.30	5536.85	6524.38
55	3294.77	3876.45	4458.13
56	4513.76	5320.43	6127.09
57	4893.48	5886.80	6880.12

EXERCISES

G3.1.
    For the data on road casualties in Table G1.5, use FDEC to find seasonality indices (note that $L = 12$ with monthly data) and to generate forecasts for the road casualties over the next 12 months (months 61–72).

G3.2.
    The number of absences at a school has been recorded on a daily basis for a five-week period, and the results are presented in Table G3.2. With FDEC, apply classical decomposition to this series, interpret the results, and find forecasts of the number of absences on the five days in the week immediately following the five-week period covered by the data in Table G3.2.

G3.3.
    What is L for the data on high temperatures in Table G1.3 and the data on inflation and unemployment in Table G1.4? Apply FDEC to determine forecasts for days 21–24 in the temperature example.

TABLE G3.2   Daily number of students absent from school for a five-week period

Day	Week 1	Week 2	Week 3	Week 4	Week 5
Monday	12	9	14	13	9
Tuesday	8	9	12	14	8
Wednesday	9	6	11	9	7
Thursday	4	7	13	7	5
Friday	7	11	18	12	10

G4

Exponential Smoothing

In using past data from a time series to forecast future values of the series, we may want to give more weight to recent observations than to data observed some time ago. Exponential smoothing is a technique that enables us to do this in a systematic manner with past values in the series being weighted in an exponentially decreasing manner. At time i, the forecast for period $i + 1$ can be written as a weighted average of the most recent observation y_i and the previous forecast $\hat{y}_i$ of y_i:

$$\hat{y}_{i+1} = \alpha y_i + (1 - \alpha)\hat{y}_i \quad , \tag{G4--1}$$

where α is a smoothing constant that determines how much the most recent observation is "smoothed". The value of α is between zero and one, with larger values indicating less smoothing. If we express past forecasts $\hat{y}_i$, $\hat{y}_{i-1}$ and so on in the form given by (G4–1), we can write $\hat{y}_{i+1}$ as a function of the previous observations:

$$\hat{y}_{i+1} = \alpha y_i + \alpha(1 - \alpha)y_{i-1} + \alpha(1 - \alpha)^2 y_{i-2}$$
$$+ \alpha(1 - \alpha)^3 y_{i-3} + \dots \quad . \tag{G4--2}$$

The weights given to the past observations decrease as we move back in time because the exponent of $1 - \alpha$, which is less than one, increases.

The basic form of exponential smoothing described above is called single exponential smoothing. Other exponential smoothing techniques have been developed. One such technique, Holt's exponential smoothing, introduces additional formulas to estimate the trend so that it can be taken into account in the forecast. For a series with a substantial trend, the inclusion of the trend in the smoothing procedure can be helpful in forecasting. Holt's exponential smoothing uses two smoothing constants, α and β, and three formulas:

$$y_i^* = \alpha y_i + (1 - \alpha)(y_{i-1}^* + T_{i-1}) \quad , \tag{G4-3}$$

$$T_i = \beta (y_1^* - y_{i-1}^*) + (1 - \beta) T_{i-1} \quad , \tag{G4-4}$$

and

$$\hat{y}_{i+k} = y_i^* + kT_i \quad . \tag{G4-5}$$

The first formula, (G4–3), is similar to (G4–1) with a term added for the trend. The trend term itself is updated in an exponential smoothing fashion in (G4–4). To generate a forecast for k time periods in the future, kT_i, a linear extrapolation of the estimated trend, is added to the current smoothed value y_i^*.

ISP COMMANDS

Single exponential smoothing can be applied via the FEXPO command in ISP. If desired, the data can be deseasonalized before exponential smoothing is conducted. The smoothing constant can be chosen by the user or determined by the program. Forecasts and actual observations can be displayed for the later observations in the time series, and forecasts are provided along with 95 percent forecast intervals for future values in the series.

The command DEXPH enables the user to apply Holt's exponential smoothing to a time series. Values of α and β can be chosen by the user or determined by the program, and the output is similar to that for DEXPO.

SOLVED EXAMPLE

For the data in Table G1.1 on quarterly sales of paper, we can use exponential smoothing to forecast future sales. For single exponential smoothing, FEXPO can generate forecasts, and some output is provided for the sales example in Figure G4.1. Since this series has a strong seasonal component, the option to deseasonalize the data is advisable, and the program computes the same seasonal indices generated by FDEC in Figure G3.1. When the program is asked to compute the smoothing constant α, this constant is chosen to maximize the accuracy of forecasts obtained for the fitting data (the data already available, consisting of quarters 1 through 51 in this example). From Figure G4.1, we see that the smoothing constant turns out to be $\alpha = 0.425$. The accuracy measures reported in the output (mean percentage error, mean absolute percentage error, root mean squared error, and R^2) provide an indication of how closely the exponential smoothing model fits the time series.

In FAUTO, forecasts and predictions for time periods near the end of the fitting data can be displayed. In Figure G4.1, this information is given for the last eight quarters.

FIGURE G4.1   Some output from FEXPO with forecasts of quarterly sales of paper generated via single exponential smoothing

***** Single Exponential Smoothing *****

What is the length of seasonality ? 4
Do you want to deseasonalize the data : Y

SEASON	SEASONAL INDICES
1	112.33
2	109.32
3	75.67
4	102.68

Do you want the program to compute the smoothing constant : Y
Value of alpha used is $= 0.425$

A C C U R A C Y Measures of M O D E L Fitting

Mean percentage error, or bias	$=$	2.35
Mean absolute percentage error	$=$	4.61
Root mean squared error	$=$	299.3
R-square	$=$	0.840

Number of the latest actual and predicted values to be printed ? 8

Period	Low Limit	F O R E C A S T	Upper Limit	Actual Value	% Error
44	4045.16	4643.70	5242.25	4525.94	−2.53
45	4426.92	5025.47	5624.01	5258.71	3.95
46	4388.50	4987.05	5585.60	5189.58	3.57
47	2913.05	3511.60	4110.15	3596.76	3.13
48	4215.68	4814.23	5412.77	3881.60	−23.40
49	4234.60	4833.15	5431.69	6231.20	19.97
50	4683.10	5281.65	5880.20	6325.40	15.09
51	3364.45	3962.99	4561.54	3602.20	−13.24

Number of forecasts ? 1

Period	Low Limit	F O R E C A S T	Upper Limit
52	4554.88	5169.49	5784.27

The forecasts are computed from (G4–1) with appropriate adjustments for seasonality. For instance, to determine a forecast for the last quarter, quarter 51, we begin by applying (G4–1) with $\alpha = 0.425$ to the observed value and forecast for quarter 50:

$$0.425(6325.40) + 0.575(5281.65) = 5725.24 \quad .$$

Quarters 50 and 51 are the second and third quarters in a year. Therefore, to adjust the forecast of 5725.24 for seasonality we divide by 1.0932 to remove the second-quarter effect and multiply by 0.7567 to include the third-quarter effect and arrive at the forecast of 3962.99 for y_{51}. From the column of percentage errors, we see that the errors are much higher for the last four quarters than for previous periods. In the last four quarters, the series took on more extreme values than was the case in the earlier portion of the series, as can be seen from the deseasonalized sales values shown in Figure G3.2.

The forecast for the next quarter, quarter 52, is computed in the same manner as the forecast for quarter 51. The weighted average

$$\alpha y_{51} + (1 - \alpha)\hat{y}_{51} = 0.425(3602.20) + 0.575(3962.99) = 3809.65$$

is divided by 0.7567 and multiplied by 1.0268 to adjust for seasonality and obtain the final forecast of 5169.49. A 95 percent forecast interval for y_{52} ranges from 4554.88 to 5784.27.

EXERCISES

G4.1.

For the temperature data in Table G1.3, use FEXPO to generate forecasts for the high temperatures on days 21 and 22, letting the program select the smoothing constant α. Then run FEXPO for the same data with $\alpha = 0.2$, $\alpha = 0.4$, $\alpha = 0.6$, and $\alpha = 0.8$. Discuss any differences among the forecasts as the smoothing constant changes.

G4.2.

Apply Holt's exponential smoothing to the temperature data in Table G1.3, using FEXPH, and compare the forecasts with those found in Exercise G4.1.

G4.3.

Use both FEXPO and FEXPH to generate forecasts for the inflation rate and the unemployment rate in year 21, given the data for years 1–20 in Table G1.4.

Summary of
Part G

An investigation of patterns in past data from a time series can be valuable in making forecasts for what might happen in the future. In turn, such forecasts are extremely useful in a wide variety of decision-making situations. In Part G, we have described methods for analyzing patterns in historical observations of variables over time and discussed some techniques for forecasting on the basis of these historical observations. Although many of the statistical methods described in earlier parts of the book can be applied to time series data, the procedures covered in Part G are specifically designed with time series data in mind.

Graphs of time series data often reveal a considerable amount of information about patterns in the series. More formal measures such as autocorrelation coefficients supplement the graphical analysis, sometimes confirming initial impressions from the graphical analysis and sometimes indicating the existence of patterns not noticed in the graphs. The decomposition of a series into components representing trend, cycle, seasonality, and randomness provides yet further information and yields specific measures (such as seasonality indices and estimates of trend) that can be used in the computation of forecasts. More sophisticated forecasting techniques than the techniques discussed here are available, but methods such as exponential smoothing are widely used and frequently perform as well as or better than some more sophisticated methods. Part G is by no means a complete development of the area of forecasting, but it does discuss some useful methods while providing an understanding of some important concepts related to time series analysis and forecasting.

VOCABULARY LIST

accuracy measures	cross-autocorrelation
autocorrelation	cycle
autocorrelation coefficient	decomposition
autocorrelation function	deseasonalized data
centered moving average	exponential smoothing
classical decomposition	first difference

forecast

forecasting

Holt's exponential smoothing

lag

length of seasonality

mean absolute percentage error

mean percentage error

medial average

moving average

randomness

root mean squared error

seasonal indices

seasonality

second difference

single exponential smoothing

smoothing constant

time series

trend

LIST OF SYMBOLS

y_i − the ith observation of y in a time series

d_i − the first difference at time i in a time series

r_k − the autocorrelation coefficient of lag k

$r_{xy}(k)$ − the cross-autocorrelation coefficient of y with x lagged k periods

T_i − the trend component in a time series

C_i − the cyclical component in a time series

S_i − the seasonality component in a time series

R_i − the randomness component in a time series

L − the length of seasonality (number of seasons) in a time series

m_i − a moving average in a time series

$\hat{y}_i$ − a forecast of y_i

α − the smoothing constant in exponential smoothing

β − the smoothing constant for updating the trend component in Holt's exponential smoothing

y_i^* − the smoothed value of y_i in Holt's exponential smoothing

FORMULAS

$$d_i \;=\; y_i - y_{i-1}$$

$$r_k \;=\; \frac{\displaystyle\sum_{i=1}^{n-k} (y_i - \bar{y})(y_{i+k} - \bar{y})}{\displaystyle\sum_{i=1}^{n} (y_i - \bar{y})^2}$$

$$r_{xy}(k) \;=\; \frac{\displaystyle\sum_{i=1}^{n-k} (x_i - \bar{x})(y_{i+k} - \bar{y})}{\sqrt{\left[\displaystyle\sum_{i=1}^{n} (x_i - \bar{x})^2\right]\left[\displaystyle\sum_{i=1}^{n} (y_i - \bar{y})^2\right]}}$$

$$y_i \;=\; T_i C_i S_i R_i$$

$$\hat{y}_{i+1} \;=\; \alpha y_i + (1 - \alpha)\hat{y}_i \qquad \text{(for single exponential smoothing)}$$

$$\hat{y}_{i+1} \;=\; \alpha y_i + \alpha(1 - \alpha)y_{i-1} + \alpha(1 - \alpha)^2 y_{i-2} + \alpha(1 - \alpha)^3 y_{i-3} + \ldots$$
$$\text{(for single exponential smoothing)}$$

$$y_i^* \;=\; \alpha y_i + (1 - \alpha)(y_{i-1}^* + T_{i-1}) \qquad \text{(for Holt's exponential smoothing)}$$

$$T_i \;=\; \beta(y_i^* - y_{i-1}^*) + (1 - \beta)T_{i-1} \qquad \text{(for Holt's exponential smoothing)}$$

$$\hat{y}_{i+k} \;=\; y_i^* + kT_i \qquad \text{(for Holt's exponential smoothing)}$$

RELEVANT ISP COMMANDS

The most relevant commands for Part G are the F commands, or forecasting commands, of ISP. FAUTO can be used to compute autocorrelation coefficients and display an autocorrelation function, either for the original data or for first or second differences.

FCROS computes cross-autocorrelation coefficients and provides a plot of these coefficients. The classical decomposition technique can be applied to decompose a time series and generate forecasts through the command FDEC. Finally, FEXPO performs single exponential smoothing and FEXPH does Holt's exponential smoothing.

Some ISP commands other than the F commands can be useful in forecasting. For example, plots of time series data generated with PLOT can be helpful in the detection of patterns in the series, and PFREQ and PHIST can provide summary measures and graphs that may be of interest. The transformation TDIF provides first and second differences from a set of data. Finally, regression is often used by itself or as part of some other procedure (for instance, to estimate the trend component of a time series) in forecasting; therefore, REGRE is valuable in forecasting.

The topics covered in Part F are explained briefly in the MF commands. These commands correspond to the four chapters in Part F.

MFG1:   Time series
MFG2:   Autocorrelation
MFG3:   Decomposition of a time series
MFG4:   Exponential smoothing

GLOSSARY

[The chapter in which a term is first discussed is indicated in parentheses.]

absolute frequency   the number of observations equal to a specified value or within a specified class of values (A1)

acceptance region   the values of a test statistic for which the null hypothesis will be accepted (D4)

accepting H_0   accepting, or failing to reject, a null hypothesis H_0 (D4)

accuracy measures   measures of the accuracy of forecasts or estimates (G4)

accuracy of estimate   the degree to which an estimate is expected to be close to the actual value of the parameter being estimated (D1)

addition rule for mutually exclusive events   the probability that one of two or more mutually exclusive events occurs is the sum of their individual probabilities (B2)

addition rule for non-mutually exclusive events   the probability that at least one of two events occurs is the sum of their individual probabilities minus the probability that they both occur (B2)

adjusted R^2   a coefficient of determination adjusted for the number of independent variables in a regression analysis (F7)

alpha   the probability of a Type I error in hypothesis testing (D4)

alternative hypothesis   a claim or hypothesis H_A that is considered as an alternative to the null hypothesis H_0 (D4)

analysis of residuals   an investigation of the assumptions of regression, involving the residuals (F5)

analysis of variance   a method for investigating the differences among several means by comparing the variability within samples and the variability between samples (E6)

analysis of variance table   a table showing the sources of variation, with sums of squares, degrees of freedom, and mean squares for an analysis of variance (E6)

area under curve as probability   the representation of probability in terms of area under a curve for continuous random variables (B9)

assumptions of linear regression   the assumptions needed to make inferences about a population regression line (F4)

autocorrelation   an interdependence between a variable and the same variable one or more periods earlier (G2)

autocorrelation coefficient   a measure of the relationship between a variable and the same variable one or more periods earlier (G2)

autocorrelation function   a graph of autocorrelations with different lags for a variable (G2)

average   a term usually indicating an arithmetic mean but sometimes used informally to refer to other measures of location (A3)

Bayes' theorem   a formula for the revision of probabilities when new information is obtained (B4)

Bayesian estimation   the process of determining point and interval estimates based on a prior or posterior distribution (D3)

Bayesian interval estimate   an interval estimate determined from a prior or posterior distribution for a parameter (D3)

Bayesian point estimate   a point estimate determined from a prior or posterior distribution for a parameter (D3)

bell-shaped distribution   a symmetric distribution shaped like a bell, such as a normal distribution (A6)

beta   the probability of a Type II error in hypothesis testing (D4)

beta distribution   a probability distribution often used to represent prior and posterior information about a proportion (D3)

binomial distribution   the probability distribution of the number of successes in n independent trials with probability p of success on any single trial (B7)

binomial probability   a probability found from a binomial distribution (B7)

bound on error of estimation   the maximum amount a point estimate is likely (with a given probability) to be in error (D1)

cell in a contingency table   a combination of particular levels of the row and column factors in a contingency table (E7)

centered moving average   a moving average centered at individual time periods (G3)

Central Limit Theorem   as the sample size increases the sampling distribution of the sample mean becomes more and more like a normal distribution (C4)

Chebyshev's theorem   a rule providing an upper bound for the relative frequency or probability within any specified number of standard deviations from the mean (A6)

chi-square distribution   a probability distribution often used for inferences about variances, goodness of fit tests, and inferences concerning contingency tables (D7)

chi-square statistic   a statistic having a chi-square distribution (D7)

class   an interval of values grouped together for the purpose of presenting or analyzing data (A1)

class interval   the width of an interval representing a class (A1)

classical decomposition   an approach often used to decompose a time series (G3)

classical interpretation of probability   an interpretation in which probabilities are based on the idea of equally likely outcomes (B1)

coding   the process of standardizing data (A7)

coefficient of determination   the square of a correlation coefficient, measuring the proportional reduction in the sum of squared errors achieved by using a regression line (F4)

coefficient of kurtosis a measure of the kurtosis, or the relative peakedness or flatness, of a distribution (A4)

coefficient of skewness a measure of the skewness, or lack of symmetry, in a distribution (A4)

coefficient of variation a measure of dispersion defined as the standard deviation divided by the mean (A4)

comparing populations the process of making inferences about differences among populations, usually in terms of specific parameters such as means, variances, or proportions (E1)

comparing several means the process of making inferences about differences among the means of several populations (E5)

comparing several proportions the process of making inferences about differences among the proportions in several populations (E7)

comparing two means the process of making inferences about differences between the means of two populations (E1)

comparing two proportions the process of making inferences about differences between the proportions in two populations (E3)

comparing two variances the process of making inferences about ratios of the variances of two populations (E4)

complementary event for event A, the complementary event is ''A does not occur'' (B2)

conditional probability the probability of an event given the occurrence of another event or series of events (B3)

confidence interval an interval of values which, with a given level of confidence, includes the parameter being estimated (D1)

confidence level the probability associated with a confidence interval (D1)

contingency table a two-dimensional array classifying items or observations according to two factors or variables (E7)

continuity correction adjustments made in the values of a random variable when using a continuous distribution to approximate a discrete distribution (such as adjustments of 0.5 when using the normal approximation to the binomial distribution) (B9)

continuous random variable a random variable with an infinite (but not countable) number of possible values (B5)

correlation an interdependence between two variables (F2)

correlation coefficient a measure of the direction and strength of the linear relationship between two variables (F2)

covariance a measure of how two variables jointly vary (F2)

critical value a value of a test statistic on the boundary between the acceptance region and the rejection region (D4)

cross-autocorrelation an interdependence between one variable and a lagged second variable (G2)

cumulative frequency the number of observations less than or equal to a specified value (A1)

cumulative histogram a graphical presentation of a cumulative frequency distribution with cumulative frequencies represented by bars, or rectangles (A2)

cumulative probability the probability that a random variable is less than or equal to a specified value (B5)

cumulative probability distribution the set of all possible values of a random variable together with a specification of the associated cumulative proababilities (B5)

cumulative relative frequency the proportion of observations less than or equal to a specified value (A1)

cycle   an effect relating to a tendency for values in a time series to show cyclical movement over time (G3)

data   a collection of measurements or observations (A1)

decision rule   in hypothesis testing, a rule indicating when to reject the null hypothesis H_0 (and therefore when to accept H_0) (D4)

decomposition   the separation of a time series into components (G3)

degrees of freedom   a parameter of the t, chi-square, and F distributions (D6)

degrees of freedom for denominator   the degrees of freedom associated with the denominator in an F statistic (E4)

degrees of freedom for numerator   the degrees of freedom associated with the numerator in an F statistic (E4)

dependent events   events that are not independent, so that the occurrence or non-occurrence of any of the events may affect the probabilities for the remaining events (B3)

dependent variable   the variable being predicted in a regression analysis (F3)

descriptive statistics   the branch of statistics concerned with describing and summarizing data (A1)

deseasonalized data   data that have been adjusted for seasonal effects by dividing each observation by the appropriate seasonal component (G3)

deviation from the mean   the difference between an individual value and the mean (A4)

difference in means   the difference between the means from two samples or populations, a measure used in the comparison of means (E1)

difference in proportions   the difference between the proportions from two samples or populations, a measure used in the comparison of proportions (E3)

discrete random variable   a random variable with a countable number of values (we can count the possible values of the variable) (B5)

dispersion   variability or ''spread'' in a set of data or a probability distribution (A4)

Durbin-Watson test   a test for the independence of errors in regression analysis (F5)

empirical rule   a rule giving approximate percentages of data within one, two, or three standard deviations from the mean for frequency distributions that are roughly bell-shaped (A6)

equally likely events   events which have the same chance of occurring (B1)

error bound   a bound on the error of estimation in using a particular sample statistic to estimate a population parameter (D1)

error of estimation   the amount an estimate differs from the actual value of the parameter being estimated (D1)

error term   the difference between an actual value and a predicted value of the dependent variable in a regression analysis (F3)

estimate   a single value or an interval of values used to approximate a population parameter (D1)

estimated regression line   an estimate of a population regression line, found from a set of data (F3)

estimation   a form of inference in which estimates are determined for population parameters of interest (D1)

estimation of a difference in means   the process of finding point and interval estimates for a difference in means (E1)

estimation of a difference in proportions   the process of finding point and interval estimates for a difference in proportions (E3)

estimation of a ratio of variances   the process of finding point and interval estimates for a ratio of variances (E4)

estimation of the mean   the process of finding point and interval estimates for a population mean (D1)

estimation of the proportion   the process of finding point and interval estimates for a population proportion (D2)

estimation of the variance   the process of finding point and interval estimates for a population variance (D7)

event   a collection of outcomes of an experiment (B2)

expected frequency   the number of times a particular event or value or class of values is expected to occur in a sample according to some stated model or hypothesis (D8)

expected value   the mean of a random variable or of some function of the random variable (B6)

experiment   any process used for obtaining a measurement or observation (B2)

explanatory variable   an independent variable (helping to explain the variation in the dependent variable) in a regression analysis (F3)

exponential function   a function involving one or more exponential terms (F1)

exponential smoothing   a forecasting technique that gives more weight to recent observations and less weight to earlier observations (G4)

F distribution   a probability distribution used for comparing two variances, for comparing several means in an analysis of variance, and for testing the goodness of fit of a regression model (E4)

F statistic   a statistic having an F distribution (E4)

finite population correction   a correction factor used to adjust the sample mean when sampling without replacement from a finite population (C3)

first difference   the difference between a variable and the same variable one time period earlier (G1)

forecast   a prediction of a future event or value of a variable (F3)

forecasting   the branch of statistics concerned with the generation of forecasts of future events or variables (G4)

frequency   the number of times an event or value or class of values occurs (A1)

frequency distribution   a listing of the frequencies for all possible values or classes of values in a set of data (A1)

function   a relationship between two variables in which the value of one variable can be determined from the value of the other variable (F1)

functional relationship   a relationship that is a function (F1)

goodness of fit   testing whether a particular model provides a ''good fit'' to a set of sample data (D8)

grand mean   the overall mean of all of the observations in a number of samples (E5)

grouped data   data which have been grouped into classes (A5)

histogram   a graphical presentation of a frequency distribution in the form of a "bar chart" with frequencies represented by bars, or rectangles (A2)

Holt's exponential smoothing   a form of exponential smoothing that takes trend into account (G4)

hypothesis   an assertion or claim about one or more populations (usually about population parameters) (D4)

hypothesis testing   a form of inference in which certain claims or hypotheses about populations are tested and accepted or rejected (D4)

hypothesis testing for a difference in means   the process of testing hypotheses about a difference in means (E1)

hypothesis testing for a difference in proportions   the process of testing hypotheses about a difference in proportions (E3)

hypothesis testing for a ratio of variances   the process of testing hypotheses about a ratio of variances (E4)

hypothesis testing for contingency tables   the process of testing hypotheses about the comparison of several proportions or about independence in contingency tables (E7)

hypothesis testing for the mean   the process of testing hypotheses about a population mean (D4)

hypothesis testing for the proportion   the process of testing hypotheses about a population proportion (D5)

hypothesis testing for the variance   the process of testing hypotheses about a population variance (D7)

hypothesis testing in analysis of variance   the process of testing for equality of means using an F statistic in the analysis of variance (E6)

independence   a situation with independent events, trials, or variables (B2)

independent events   events are independent if the occurrence or non-occurrence of any of the events has no bearing on the probabilities for the remaining events (B2)

independent samples   two or more samples that are independent of each other (E1)

independent trials   a series of trials for which the outcome on any trial is independent of the outcomes on the other trials (B7)

independent variable   a variable used to help predict a dependent variable in a regression analysis (F3)

independent variables   variables are independent if the values of any of the variables have no bearing on the probability distributions for the remaining variables (F2)

inference   the process of reasoning from sample data to an entire population; a statement about a population based on a sample (C5)

inferences from large samples   inferences about a population based on large samples (for example, samples large enough for the Central Limit Theorem to be applicable) (D1)

inferences from small samples   inferences about a population based on small samples (for example, samples not large enough for the Central Limit Theorem to be applicable) (D6)

inferential statistics   the branch of statistics concerned with the use of information from a sample to say something about the population from which the sample was drawn (C5)

intercept of regression line   the predicted value of the dependent variable in a regression analysis when all of the independent variables are zero (F3)

interval estimate   a range of values used to estimate a population parameter (D1)

joint probability   for two events, the probability that both occur (B2)

kurtosis   the relative peakedness or flatness of a frequency or probability distribution (A4)

lag   the consideration of a variable a number of periods earlier (G2)

law of large numbers   the relative frequency of occurrence of an event in a series of independent trials tends to get closer and closer to the probability of the event as the number of trials increases (B1)

least squares   a method for choosing a line or curve to fit a set of data by minimizing the sum of the squared errors (F3)

length of seasonality   the number of seasons in time series data (G3)

likelihood   the probability that certain new information would occur given a particular event (B4)

linear function   a function in which each variable is simply given to the first degree (with two variables, a function that can be represented on a graph by a straight line) (F1)

linear regression   regression analysis with the dependent variable expressed as a linear function of the independent variables (F3)

location   the notion of a "typical" value in a set of data or a probability distribution (A3)

logarithmic function   a function involving one or more logarithmic terms (F1)

marginal probability   the probability of an event, not conditional on any other event (B3)

mean   a measure of location defined as an arithmetic average of a set of data or an expected value of a random variable (A3)

mean absolute percentage error   the average absolute percentage error in a set of forecasts (G4)

mean percentage error   the average percentage error in a set of forecasts (G4)

mean square between groups   the sum of squares between groups divided by the degrees of freedom between groups (E6)

mean square within groups   the sum of squares within groups divided by the degrees of freedom within groups (E6)

measure of dispersion   a measure indicating how "spread out" a set of data or a probability distribution is (A4)

measure of location   a measure indicating an "average" or "typical" member of a data set or "typical" value in a probability distribution (A3)

medial average   an average calculated after extreme values are discarded (G3)

median   a measure of location defined as a "middle value" in a set of data or a probability distribution (A3)

midpoint   the value midway between the lower and upper limits of a class or interval (A1)

mode   a measure of location defined as the value with the highest frequency in a set of data or the highest probability in a probability distribution (A3)

moving average   an average of a number of successive observations in a time series (G3)

multiple regression   regression analysis (usually linear regression) with more than one independent variable (F7)

multiplication rule for dependent events   for two events, the probability of both occurring is the probability of the first event multiplied by the conditional probability of the second event given that the first event occurs (B3)

multiplication rule for independent events   for two independent events, the probability of both occurring is the product of their individual probabilities (B2)

mutually exclusive events   events are mutually exclusive if at most one of the events can occur (the occurrence of one precludes the occurrence of any of the others) (B2)

negative relationship   a relationship in which one variable tends to move in the opposite direction from a second variable (when one increases, the other tends to decrease) (F2)

nonlinear relationship   a relationship in which one variable is expressed as a nonlinear function of one or more other variables (F6)

normal distribution   a widely-used continuous probability distribution with a symmetric bell-shaped curve (B9)

normal probability   a probability found from a normal distribution (B9)

null hypothesis   a claim or hypothesis H_0 that is to be tested against some alternative hypothesis (D4)

observed frequency   the number of times a particular event or value or class of values actually occurs in a sample (D8)

odds   the relative chance of an event occurring (the probability that the event occurs divided by the probability that the event does not occur) (B1)

one-tailed test (one-sided test) to the left   a test in which the alternative hypothesis includes only values of the parameter of interest less than the value specified in the null hypothesis (D4)

one-tailed test (one-sided test) to the right   a test in which the alternative hypothesis includes only values of the parameter of interest greater than the value specified in the null hypothesis (D4)

ordered data   data arranged in order of magnitude (A1)

outcome   a specific result of an experiment (B2)

paired samples   two samples that are paired in the sense that each observation from one sample is paired in some way with a single observation from the other sample (E1)

pairs of observations   a set of data consisting of pairs of values of two variables (F2)

parameter   a summary measure for a population or a probability distribution (A4)

pie chart   a graphical presentation of a frequency distribution with frequencies represented by sectors of a circle (A2)

point estimate   a single number used to estimate a population parameter (D1)

Poisson distribution   the probability distribution of the number of occurrences of an event over a continuum such as time when the occurrences are independent and occur at an average rate of λ (B8)

Poisson probability   a probability found from a Poisson distribution (B8)

pooled estimate of a proportion   an estimate obtained by pooling two sample proportions from two separate samples (E3)

pooled estimate of a standard deviation   an estimate obtained by pooling two standard deviations from two separate samples (E2)

population   the collection of the elements or observations that are of interest in a particular problem (A4)

population mean   the mean of a probability distribution or of a set of data representing a population (A4)

population proportion   the relative frequency or the probability of occurrence of an event of interest in a population (D2)

population regression line   a line representing the relationship between an independent variable and a dependent variable in the population (F4)

population standard deviation   the standard deviation of a probability distribution or a set of data representing a population (A4)

population variance   the variance of a probability distribution or a set of data representing a population (A4)

positive relationship   a relationship in which one variable tends to move in the same direction as a second variable (when one increases, the other tends to increase) (F2)

posterior distribution   a probability distribution representing information available after a sample is taken (D3)

posterior information   the information known after a sample is taken (prior information combined with sample information) (D3)

posterior probability   the probability of an event after new information is seen (B4)

prediction   a forecast, often a forecast of a dependent variable from a regression analysis (F3)

prior distribution   a probability distribution representing information available before a sample is taken (D3)

prior information   the information known before a sample is taken (D3)

prior probability   the probability of an event before new information is seen (B4)

probability   a numerical measure indicating how likely an event is to occur (B1)

probability distribution   the set of all possible values of a random variable together with a specification of the associated probabilities (B5)

probability sampling   a process of choosing a sample from a population such that each possible sample has a known probability of being selected (C1)

proportion   the relative frequency or probability of occurrence of an event of interest (D2)

quadratic function   a polynomial function of degree two (involving a variable squared but not raised to a higher power) (F1)

random numbers   a set of numbers chosen in such a way that all possible numbers are equally likely to be selected and the order of the numbers is random (C1)

random sample   a sample chosen by a random selection process (C1)

random variable   a numerical quantity, the value of which we are uncertain about (more formally, a well-defined rule for assigning a numerical value to each possible outcome of an experiment) (B5)

randomness   variation in a set of data not attributable to any systematic factors (G3)

range   a measure of dispersion defined as the difference between the largest and smallest values in a set of data or a probability distribution (A4)

ranked data   data arranged in order of magnitude (A1)

ratio of variances   a ratio of variances from two different samples or populations, a measure used in the comparison of variances (E4)

regression   a model in which a prediction of one variable is expressed as a function of one or more other variables (F3)

regression coefficients   the coefficients in a regression equation (F3)

regression line   a line expressing one variable as a linear function of another variable (F3)

rejecting H_0   rejecting a null hypothesis H_0 in favor of an alternative hypothesis H_A (D4)

rejection region   the values of a test statistic for which the null hypothesis will be rejected in favor of the alternative hypothesis (D4)

relative frequency   the proportion of observations equal to a specified value or within a specified class of values (A1)

relative frequency interpretation of probability   an interpretation in which probabilities are based on empirical evidence (relative frequencies of occurrence in a long series of independent trials) (B1)

representative sample   a sample which is representative of the population from which it was drawn (exhibits the essential characteristics of the population) (C1)

residual   the difference between an actual value and a predicted value of the dependent variable in an estimated regression line (F5)

revision of probabilities   the use of Bayes' theorem to adjust probabilities appropriately as new information is obtained (B4)

root mean squared error   the square root of the average squared error in a set of forecasts (G4)

rules of probability   rules or formulas which relate certain probabilities to each other and are therefore useful in the calculation of probabilities (B2)

sample   a subset, or portion, of the elements or observations in a population (A4)

sample mean   the arithmetic mean of a set of data representing a sample (A4)

sample proportion   the relative frequency of occurrence of an event of interest in a sample (D2)

sample size   the number of observations in a sample (A4)

sample space   all possible outcomes of an experiment (B2)

sample standard deviation   the standard deviation of a set of data representing a sample (A4)

sample variance   the variance of a set of data representing a sample (A4)

sampling   the process of choosing a sample from a population (C1)

sampling distribution   a probability distribution of a sample statistic (C2)

sampling distribution of the mean   the probability distribution of the sample mean (C3)

sampling fluctuations   differences among samples due to chance (C2)

sampling-theory estimation (classical estimation)   the process of determining point and interval estimates based on the information from a sample (D1)

sampling without replacement   a sampling plan in which a member of the population, when selected for the sample, is not replaced in the population and therefore cannot be included in the sample more than once (C1)

sampling with replacement   a sampling plan in which a member of the population, when selected for the sample, is replaced in the population (C1)

scatter diagram   a two-dimensional graph with points representing pairs of values of two variables (F1)

seasonal indices   measures indicating the effects of different seasons in a time series (G3)

seasonality   an effect relating to seasons in a time series (G3)

second difference   the difference between successive first differences in a time series (G1)

simple linear regression   linear regression with only one independent variable (F3)

simple random sampling   a sampling plan in which every possible sample of n members of the population has the same chance of being selected (C1)

single exponential smoothing   a simple form of exponential smoothing for forecasting (G4)

slope of regression line   the rate at which the predicted value from a regression line changes as the independent variable increases by one unit (F3)

skewed distribution   a frequency distribution or probability distribution that is not symmetric (A4)

skewness   the lack of symmetry in a frequency or probability distribution (A4)

smoothing constant   a value that determines how much weight is given to recent observations and earlier observations in exponential smoothing (G4)

standard deviation   a measure of dispersion defined as the square root of the variance (A4)

standard deviation of prediction error   the standard deviation of the error in a prediction from a regression analysis (F4)

standard deviation of the regression   an estimate of the standard deviation of the error term in a regression analysis (F4)

standard error   the standard deviation of the sampling distribution of a statistic (C3)

standard error of a difference in means   the standard deviation of the sampling distribution of a difference in sample means (E1)

standard error of a difference in proportions   the standard deviation of the sampling distribution of a difference in sample proportions (E3)

standard error of a regression coefficient   the standard deviation of the sampling distribution of a regression coefficient (F4)

standard error of the mean   the standard deviation of the sampling distribution of the sample mean (C3)

standard error of the sample proportion   the standard deviation of the sampling distribution of the sample proportion (D2)

standard normal distribution   a normal distribution with mean zero and standard deviation one (B9)

standardized values   values obtained by subtracting the mean from the original values and then dividing by the standard deviation (A7)

standardizing data   converting to standardized values by subtracting the mean from the original values and then dividing by the standard deviation (A7)

statistical inference   the process of using information from a sample to say something about the population from which the sample was drawn (C5)

statistic   a summary measure or other quantity calculated from a sample (A4)

statistics   a field of study involving the collection and analysis of data, including making inferences and decisions as well as simply describing and summarizing data (A1)

stratified random sampling   a sampling plan in which the population is divided into groups (strata) and separate random samples are taken from the groups (C1)

strength of relationship   the degree to which the relationship between two variables follows a particular function (F1)

subjective interpretation of probability   an interpretation in which probabilities are based on an individual's subjective judgments (B1)

sum of squares between groups   the sum of the squared deviations of the individual sample means from the grand sample mean, with each squared deviation multiplied by the appropriate sample size, in an analysis of variance (E6)

sum of squares within groups   the sum of the squared deviations of the observations from their respective sample means in an analysis of variance (E6)

summary statistic   a measure of some characteristic (such as location or dispersion) for a sample (A4)

symmetric distribution   a frequency distribution or probability distribution that is symmetric in the sense that the left side of the distribution is a mirror image of the right side (A4)

systematic bias   in sampling, a systematic tendency for a sampling plan to result in a sample that is not representative of the population (C1)

t distribution   a probability distribution often used for small-sample inferences about means and inferences about parameters in regression and correlation problems (D6)

t statistic   a statistic having a t distribution (D6)

test of independence in contingency tables   a test (using a chi-square statistic) of the independence of the row and column factors or variables in a contingency table (E7)

test statistic   a statistic on which the decision to accept a null hypothesis or reject it in favor of an alternative hypothesis is based (D4)

time series   observations made on a variable over time (G1)

total sum of squares   the sum of the squared deviations of the observations from the grand sample mean in an analysis of variance (E6)

transformation   the process of transforming one variable into another variable which is a function of the first variable, often used to deal with nonlinear relationships in regression analysis (F6)

treatment effects   the average effects of given treatments, or groups, or populations, measured by differences between individual means and the grand mean, in comparing several populations (E5)

trend   an effect relating to a tendency for values in a time series to increase or decrease over time (G3)

trial   one of a series of repeated experiments (B7)

two-tailed test (two-sided test)   a test in which the alternative hypothesis includes values of the parameter of interest on both sides of the value specified in the null hypothesis (D4)

Type I error   rejecting a null hypothesis when it is in fact true (D4)

Type II error   accepting a null hypothesis when it is actually false and the alternative hypothesis is true (D4)

unbiased estimate   the value of a statistic having a sampling distribution with a mean equal to the parameter being estimated (D1)

uncertainty   the state of not being sure about the occurrence of an event or about the value of a variable (B1)

variability between samples   differences between samples, or between groups, attributable to both sampling fluctuations and differences in the population means, in a comparison of means with the analysis of variance (E6)

variability within samples   differences within samples, or within groups, attributable to sampling fluctuations, in a comparison of means with the analysis of variance (E6)

variance   a measure of dispersion involving an average or expected value of squared deviations from the mean (A4)

z-scores (z-values)   standardized values obtained by subtracting the mean from the original values and then dividing by the standard deviation (A7)

z statistic   a statistic having a standard normal distribution (D1)

Tables

For convenience, some tables involving frequently-used probability distributions are presented here. The tables are as follows:

Table T1—A table of right-tail standard normal probabilities for certain z-values
Table T2—A table of t-values for certain specified right-tail probabilities
Table T3—A table of chi-square values for certain right-tail probabilities
Table T4—A table of F-values for a right-tail probability of 0.05
Table T5—A table of binomial probabilities for selected values of n and p
Table T6—A table of Poisson probabilities for selected values of λ

For other probabilities, such as binomial probabilities for values of n and p not given in Table T5, use the appropriate ISP commands (for instance, DBINO in the case of binomial probabilities).

TABLE T1 A table of right-tail standard normal probabilities for certain z-values

z	.00	.01	.02	.03	.04	.05	.06	.07	.08	.09
0.0	.5000	.4960	.4920	.4880	.4840	.4801	.4761	.4721	.4681	.4641
.1	.4602	.4562	.4522	.4483	.4443	.4404	.4364	.4325	.4286	.4247
.2	.4207	.4168	.4129	.4090	.4052	.4013	.3974	.3936	.3897	.3859
.3	.3821	.3783	.3745	.3707	.3669	.3632	.3594	.3557	.3520	.3483
.4	.3446	.3409	.3372	.3336	.3300	.3264	.3228	.3192	.3156	.3121
.5	.3085	.3050	.3015	.2981	.2946	.2912	.2877	.2843	.2810	.2776
.6	.2743	.2709	.2676	.2643	.2611	.2578	.2546	.2514	.2483	.2451
.7	.2420	.2389	.2358	.2327	.2296	.2266	.2236	.2206	.2177	.2148
.8	.2119	.2090	.2061	.2033	.2005	.1977	.1949	.1922	.1894	.1817
.9	.1841	.1814	.1788	.1762	.1736	.1711	.1685	.1660	.1635	.1611
1.0	.1587	.1562	.1539	.1515	.1492	.1469	.1446	.1423	.1401	.1379
1.1	.1357	.1335	.1314	.1292	.1271	.1251	.1230	.1210	.1190	.1170
1.2	.1151	.1131	.1112	.1093	.1075	.1056	.1038	.1020	.1003	.0985
1.3	.0968	.0951	.0934	.0918	.0901	.0885	.0869	.0853	.0838	.0823
1.4	.0808	.0793	.0778	.0764	.0749	.0735	.0721	.0708	.0694	.0681
1.5	.0668	.0655	.0643	.0630	.0618	.0606	.0594	.0582	.0571	.0559
1.6	.0548	.0537	.0526	.0516	.0505	.0495	.0485	.0475	.0465	.0455
1.7	.0446	.0436	.0427	.0418	.0409	.0401	.0392	.0384	.0375	.0367
1.8	.0359	.0351	.0344	.0336	.0329	.0322	.0314	.0307	.0301	.0294
1.9	.0287	.0281	.0274	.0268	.0262	.0256	.0250	.0244	.0239	.0233
2.0	.0228	.0222	.0217	.0212	.0207	.0202	.0197	.0192	.0188	.0183
2.1	.0179	.0174	.0170	.0166	.0162	.0158	.0154	.0150	.0146	.0143
2.2	.0139	.0136	.0132	.0129	.0125	.0122	.0119	.0116	.0113	.0110
2.3	.0107	.0104	.0102	.0099	.0096	.0094	.0091	.0089	.0087	.0084
2.4	.0082	.0080	.0078	.0075	.0073	.0071	.0069	.0068	.0066	.0064
2.5	.0062	.0060	.0059	.0057	.0055	.0054	.0052	.0051	.0049	.0049
2.6	.0047	.0045	.0044	.0043	.0041	.0040	.0039	.0038	.0037	.0036
2.7	.0035	.0034	.0033	.0032	.0031	.0030	.0029	.0028	.0027	.0026
2.8	.0026	.0025	.0024	.0023	.0023	.0022	.0021	.0021	.0020	.0019
2.9	.0019	.0018	.0018	.0017	.0016	.0016	.0015	.0015	.0014	.0014
3.0	.0013	.0013	.0013	.0012	.0012	.0011	.0011	.0011	.0010	.0010

Example: $P(z \geq 1.96) = 0.0250$

TABLE T2   A table of *t*-values for certain specified right-tail probabilities

d.f.	.100	.050	.025	.010	.005
1	3.078	6.314	12.706	31.821	63.657
2	1.886	2.920	4.303	6.965	9.925
3	1.638	2.353	3.182	4.541	5.841
4	1.533	2.132	2.776	3.747	4.604
5	1.476	2.015	2.571	3.365	4.032
6	1.440	1.943	2.447	3.143	3.707
7	1.415	1.895	2.365	2.998	3.499
8	1.397	1.860	2.306	2.896	3.355
9	1.383	1.833	2.262	2.821	3.250
10	1.372	1.812	2.228	2.764	3.169
11	1.363	1.796	2.201	2.718	3.106
12	1.356	1.782	2.179	2.681	3.055
13	1.350	1.771	2.160	2.650	3.012
14	1.345	1.761	2.145	2.624	2.977
15	1.341	1.753	2.131	2.602	2.947
16	1.337	1.746	2.120	2.583	2.921
17	1.333	1.740	2.110	2.567	2.898
18	1.330	1.734	2.101	2.552	2.878
19	1.328	1.729	2.093	2.539	2.861
20	1.325	1.725	2.086	2.528	2.845
21	1.323	1.721	2.080	2.518	2.831
22	1.321	1.717	2.074	2.508	2.819
23	1.319	1.714	2.069	2.500	2.807
24	1.318	1.711	2.064	2.492	2.797
25	1.316	1.708	2.060	2.485	2.787
26	1.315	1.706	2.056	2.479	2.779
27	1.314	1.703	2.052	2.473	2.771
28	1.313	1.701	2.048	2.467	2.763
29	1.311	1.699	2.045	2.462	2.756

Example: With 14 degrees of freedom, $P(t \geq 1.761) = 0.05$

TABLE T3   A table of chi-square values for certain right-tail probabilities

d.f.	.995	.990	.975	.950	.050	.025	.010	.005
1	0.000	0.000	0.001	0.004	3.841	5.024	6.635	7.879
2	0.010	0.020	0.051	0.103	5.991	7.378	9.210	10.597
3	0.072	0.115	0.216	0.352	7.815	9.348	11.345	12.838
4	0.207	0.297	0.484	0.711	9.488	11.143	13.277	14.860
5	0.412	0.554	0.831	1.145	11.070	12.832	15.086	16.750
6	0.676	0.872	1.237	1.635	12.592	14.449	16.812	18.548
7	0.989	1.239	1.690	2.167	14.067	16.013	18.475	20.278
8	1.344	1.646	2.180	2.733	15.507	17.535	20.090	21.955
9	1.735	2.088	2.700	3.325	16.919	19.023	21.666	23.589
10	2.156	2.558	3.247	3.940	18.307	20.483	23.209	25.188
11	2.603	3.053	3.816	4.575	19.675	21.920	24.725	26.757
12	3.074	3.571	4.404	5.226	21.026	23.337	26.217	28.300
13	3.565	4.107	5.009	5.892	22.362	24.736	27.688	29.819
14	4.075	4.660	5.629	6.571	23.685	26.119	29.141	31.319
15	4.601	5.229	6.262	7.261	24.996	27.488	30.578	32.801
16	5.142	5.812	6.908	7.962	26.296	28.845	32.000	34.267
17	5.697	6.408	7.564	8.672	27.587	30.191	33.409	35.718
18	6.265	7.015	8.231	9.390	28.869	31.526	34.805	37.156
19	6.844	7.633	8.907	10.117	30.144	32.852	36.191	38.582
20	7.434	8.260	9.591	10.851	31.410	34.170	37.566	39.997
21	8.034	8.897	10.283	11.591	32.671	35.479	38.932	41.401
22	8.643	9.542	10.982	12.338	33.924	36.781	40.289	42.796
23	9.260	10.196	11.689	13.091	35.172	38.076	41.638	44.181
24	9.886	10.856	12.401	13.848	36.415	39.364	42.980	45.558
25	10.520	11.524	13.120	14.611	37.652	40.646	44.314	46.928
26	11.160	12.198	13.844	15.379	38.885	41.923	45.642	48.290
27	11.808	12.879	14.573	16.151	40.113	43.194	46.963	49.645
28	12.461	13.565	15.308	16.928	41.337	44.461	48.278	50.993
29	13.121	14.256	16.047	17.708	42.557	45.722	49.588	52.336
30	13.787	14.953	16.791	18.493	43.773	46.979	50.892	53.672

Example: With 9 degrees of freedom, $P(\chi^2 \geq 16.919) = 0.05$

TABLE T4 A table of F-values for a right-tail probability of 0.05

D.f. for denom.	Degrees of freedom for numerator																		
	1	2	3	4	5	6	7	8	9	10	12	15	20	24	30	40	60	120	∞
1	161.4	199.5	215.7	224.6	230.2	234.0	236.8	238.9	240.5	241.9	243.9	245.9	248.0	249.1	250.1	251.1	252.2	253.3	254.3
2	18.51	19.00	19.16	19.25	19.30	19.33	19.35	19.37	19.38	19.40	19.41	19.43	19.45	19.45	19.46	19.47	19.48	19.49	19.50
3	10.13	9.55	9.28	9.12	9.01	8.94	8.89	8.85	8.81	8.79	8.74	8.70	8.66	8.64	8.62	8.59	8.57	8.55	8.53
4	7.71	6.94	6.59	6.39	6.26	6.16	6.09	6.04	6.00	5.96	5.91	5.86	5.80	5.77	5.75	5.72	5.69	5.66	5.63
5	6.61	5.79	5.41	5.19	5.05	4.95	4.88	4.82	4.77	4.74	4.68	4.62	4.56	4.53	4.50	4.46	4.43	4.40	4.36
6	5.99	5.14	4.76	4.53	4.39	4.28	4.21	4.15	4.10	4.06	4.00	3.94	3.87	3.84	3.81	3.77	3.74	3.70	3.67
7	5.59	4.74	4.35	4.12	3.97	3.87	3.79	3.73	3.68	3.64	3.57	3.51	3.44	3.41	3.38	3.34	3.30	3.27	3.23
8	5.32	4.46	4.07	3.84	3.69	3.58	3.50	3.44	3.39	3.35	3.28	3.22	3.15	3.12	3.08	3.04	3.01	2.97	2.93
9	5.12	4.26	3.86	3.63	3.48	3.37	3.29	3.23	3.18	3.14	3.07	3.01	2.94	2.90	2.86	2.83	2.79	2.75	2.71
10	4.96	4.10	3.71	3.48	3.33	3.22	3.14	3.07	3.02	2.98	2.91	2.85	2.77	2.74	2.70	2.66	2.62	2.58	2.54
11	4.84	3.98	3.59	3.36	3.20	3.09	3.01	2.95	2.90	2.85	2.79	2.72	2.65	2.61	2.57	2.53	2.49	2.45	2.40
12	4.75	3.89	3.49	3.26	3.11	3.00	2.91	2.85	2.80	2.75	2.69	2.62	2.54	2.51	2.47	2.43	2.38	2.34	2.30
13	4.67	3.81	3.41	3.18	3.03	2.92	2.83	2.77	2.71	2.67	2.60	2.53	2.46	2.42	2.38	2.34	2.30	2.25	2.21
14	4.60	3.74	3.34	3.11	2.96	2.85	2.76	2.70	2.65	2.60	2.53	2.46	2.39	2.35	2.31	2.27	2.22	2.18	2.13
15	4.54	3.68	3.29	3.06	2.90	2.79	2.71	2.64	2.59	2.54	2.48	2.40	2.33	2.29	2.25	2.20	2.16	2.11	2.07
16	4.49	3.63	3.24	3.01	2.85	2.74	2.66	2.59	2.54	2.49	2.42	2.35	2.28	2.24	2.19	2.15	2.11	2.06	2.01
17	4.45	3.59	3.20	2.96	2.81	2.70	2.61	2.55	2.49	2.45	2.38	2.31	2.23	2.19	2.15	2.10	2.06	2.01	1.96
18	4.41	3.55	3.16	2.93	2.77	2.66	2.58	2.51	2.46	2.41	2.34	2.27	2.19	2.15	2.11	2.06	2.02	1.97	1.92
19	4.38	3.52	3.13	2.90	2.74	2.63	2.54	2.48	2.42	2.38	2.31	2.23	2.16	2.11	2.07	2.03	1.98	1.93	1.88
20	4.35	3.49	3.10	2.87	2.71	2.60	2.51	2.45	2.39	2.35	2.28	2.20	2.12	2.08	2.04	1.99	1.95	1.90	1.84
21	4.32	3.47	3.07	2.84	2.68	2.57	2.49	2.42	2.37	2.32	2.25	2.18	2.10	2.05	2.01	1.96	1.92	1.87	1.81
22	4.30	3.44	3.05	2.82	2.66	2.55	2.46	2.40	2.34	2.30	2.23	2.15	2.07	2.03	1.98	1.94	1.89	1.84	1.78
23	4.28	3.42	3.03	2.80	2.64	2.53	2.44	2.37	2.32	2.27	2.20	2.13	2.05	2.01	1.96	1.91	1.86	1.81	1.76
24	4.26	3.40	3.01	2.78	2.62	2.51	2.42	2.36	2.30	2.25	2.18	2.11	2.03	1.98	1.94	1.89	1.84	1.79	1.73
25	4.24	3.39	2.99	2.76	2.60	2.49	2.40	2.34	2.28	2.24	2.16	2.09	2.01	1.96	1.92	1.87	1.82	1.77	1.71
26	4.23	3.37	2.98	2.74	2.59	2.47	2.39	2.32	2.27	2.22	2.15	2.07	1.99	1.95	1.90	1.85	1.80	1.75	1.69
27	4.21	3.35	2.96	2.73	2.57	2.46	2.37	2.31	2.25	2.20	2.13	2.06	1.97	1.93	1.88	1.84	1.79	1.73	1.67
28	4.20	3.34	2.95	2.71	2.56	2.45	2.36	2.29	2.24	2.19	2.12	2.04	1.96	1.91	1.87	1.82	1.77	1.71	1.65
29	4.18	3.33	2.93	2.70	2.55	2.43	2.35	2.28	2.22	2.18	2.10	2.03	1.94	1.90	1.85	1.81	1.75	1.70	1.64
30	4.17	3.32	2.92	2.69	2.53	2.42	2.33	2.27	2.21	2.16	2.09	2.01	1.93	1.89	1.84	1.79	1.74	1.68	1.62
40	4.08	3.23	2.84	2.61	2.45	2.34	2.25	2.18	2.12	2.08	2.00	1.92	1.84	1.79	1.74	1.69	1.64	1.58	1.51
60	4.00	3.15	2.76	2.53	2.37	2.25	2.17	2.10	2.04	1.99	1.92	1.84	1.75	1.70	1.65	1.59	1.53	1.47	1.39
120	3.92	3.07	2.68	2.45	2.29	2.17	2.09	2.02	1.96	1.91	1.83	1.75	1.66	1.61	1.55	1.50	1.43	1.35	1.25
∞	3.84	3.00	2.60	2.37	2.21	2.10	2.01	1.94	1.88	1.83	1.75	1.67	1.57	1.52	1.46	1.39	1.32	1.22	1.00

Example: With 10 d.f. for the numerator and 15 d.f. for the denominator, $P(F \geq 2.54) = 0.05$.

TABLE T5   A table of binomial probabilities for selected values of n and p

n	x	.10	.20	p .30	.40	.50
1	0	.9000	.8000	.7000	.6000	.5000
	1	.1000	.2000	.3000	.4000	.5000
2	0	.8100	.6400	.4900	.3600	.2500
	1	.1800	.3200	.4200	.4800	.5000
	2	.0100	.0400	.0900	.1600	.2500
3	0	.7290	.5120	.3430	.2160	.1250
	1	.2430	.3840	.4410	.4320	.3750
	2	.0270	.0960	.1890	.2880	.3750
	3	.0010	.0080	.0270	.0640	.1250
4	0	.6561	.4096	.2401	.1296	.0625
	1	.2916	.4096	.4116	.3456	.2500
	2	.0486	.1536	.2646	.3456	.3750
	3	.0036	.0256	.0756	.1536	.2500
	4	.0001	.0016	.0081	.0256	.0625
5	0	.5905	.3277	.1681	.0078	.0312
	1	.3280	.4096	.3602	.2592	.1562
	2	.0729	.2048	.3087	.3456	.3125
	3	.0081	.0512	.1323	.2304	.3125
	4	.0004	.0064	.0284	.0768	.1562
	5	.0000	.0003	.0024	.0102	.0312
6	0	.5314	.2621	.1176	.0467	.0156
	1	.3543	.3932	.3025	.1866	.0938
	2	.0984	.2458	.324\	.3110	.2344
	3	.0146	.0819	.1852	.2765	.3125
	4	.0012	.0154	.0595	.1382	.2344
	5	.0001	.0015	.0102	.0369	.0938
	6	.0000	.0001	.0007	.0041	.0156
7	0	.4783	.2097	.0824	.0280	.0078
	1	.3720	.3670	.2471	.1306	.0547
	2	.1240	.2753	.3177	.2613	.1641
	3	.0230	.1147	.2269	.2903	.2734
	4	.0026	.0287	.0972	.1935	.2734
	5	.0002	.0043	.0250	.0774	.1641
	6	.0000	.0004	.0036	.0172	.0547
	7	.0000	.0000	.0002	.0016	.0078

TABLE T5 A table of binomial probabilities for selected values of *n* and *p* (cont.)

n	x	.10	.20	p .30	.40	.50
8	0	.4305	.1678	.0576	.0168	.0039
	1	.3826	.3355	.1977	.0896	.0312
	2	.1488	.2936	.2965	.2090	.1094
	3	.0331	.1468	.2541	.2787	.2188
	4	.0046	.0459	.1361	.2322	.2734
	5	.0004	.0092	.0467	.1239	.2188
	6	.0000	.0011	.0100	.0413	.1094
	7	.0000	.0001	.0012	.0079	.0312
	8	.0000	.0000	.0001	.0007	.0039
9	0	.3874	.1342	.0404	.0101	.0020
	1	.3874	.3020	.1556	.0605	.0176
	2	.1722	.3020	.2668	.1612	.0703
	3	.0446	.1762	.2668	.2508	.1641
	4	.0074	.0661	.1715	.2508	.2461
	5	.0008	.0165	.0735	.1672	.2461
	6	.0001	.0028	.0210	.0743	.1641
	7	.0000	.0003	.0039	.0212	.0703
	8	.0000	.0000	.0004	.0035	.0176
	9	.0000	.0000	.0000	.0003	.0020
10	0	.3487	.1074	.0282	.0060	.0010
	1	.3874	.2684	.1211	.0403	.0098
	2	.1937	.3020	.2335	.1209	.0439
	3	.0574	.2013	.2668	.2150	.1172
	4	.0112	.0881	.2001	.2508	.2051
	5	.0015	.0264	.1029	.2007	.2461
	6	.0001	.0055	.0368	.1115	.2051
	7	.0000	.0008	.0090	.0425	.1172
	8	.0000	.0001	.0014	.0106	.0439
	9	.0000	.0000	.0001	.0016	.0098
	10	.0000	.0000	.0000	.0001	.0010
11	0	.3138	.0859	.0198	.0036	.0005
	1	.3835	.2362	.0932	.0266	.0054
	2	.2131	.2953	.1998	.0887	.0269
	3	.0710	.2215	.2568	.1774	.0806
	4	.0158	.1107	.2201	.2365	.1611
	5	.0025	.0388	.1231	.2207	.2256
	6	.0003	.0097	.0566	.1471	.2256
	7	.0000	.0017	.0173	.0701	.1611
	8	.0000	.0002	.0037	.0234	.0806
	9	.0000	.0000	.0005	.0052	.0269
	10	.0000	.0000	.0000	.0007	.0054
	11	.0000	.0000	.0000	.0000	.0005

TABLE T5   A table of binomial probabilities for selected values of *n* and *p* (cont.)

n	x	.10	.20	$\frac{p}{.30}$	.40	.50
12	0	.2824	.0687	.0138	.0022	.0002
	1	.3766	.2062	.0712	.0174	.0029
	2	.2301	.2835	.1678	.0639	.0161
	3	.0852	.2362	.2397	.1419	.0537
	4	.0213	.1329	.2311	.2128	.1208
	5	.0038	.0532	.1585	.2270	.1934
	6	.0005	.0155	.0792	.1766	.2256
	7	.0033	.0000	.0291	.1009	.1934
	8	.0005	.0000	.0078	.0420	.1208
	9	.0001	.0000	.0015	.0125	.0537
	10	.0000	.0000	.0002	.0025	.0161
	11	.0000	.0000	.0000	.0003	.0029
	12	.0000	.0000	.0000	.0000	.0002
13	0	.0550	.2542	.0097	.0013	.0001
	1	.1787	.3672	.0540	.0113	.0016
	2	.2680	.2448	.1388	.0453	.0095
	3	.2457	.0997	.2181	.1107	.0349
	4	.1535	.0277	.2337	.1845	.0873
	5	.0691	.0055	.1803	.2214	.1571
	6	.0230	.0008	.1030	.1968	.2095
	7	.0058	.0001	.0442	.1312	.2095
	8	.0011	.0000	.0142	.0656	.1571
	9	.0001	.0000	.0034	.0243	.0873
	10	.0000	.0000	.0006	.0065	.0349
	11	.0000	.0000	.0001	.0012	.0095
	12	.0000	.0000	.0000	.0001	.0016
	13	.0000	.0000	.0000	.0000	.0001
14	0	.0440	.2288	.0068	.0008	.0001
	1	.1539	.3559	.0407	.0073	.0009
	2	.2501	.2570	.1134	.0317	.0056
	3	.2501	.1142	.1943	.0845	.0222
	4	.1720	.0349	.2290	.1549	.0611
	5	.0860	.0078	.1963	.2066	.1222
	6	.0322	.0013	.1262	.2066	.1833
	7	.0092	.0002	.0618	.1574	.2095
	8	.0020	.0000	.0232	.0918	.1833
	9	.0003	.0000	.0066	.0408	.1222
	10	.0000	.0000	.0014	.0136	.0611
	11	.0000	.0000	.0002	.0033	.0222
	12	.0000	.0000	.0000	.0005	.0056
	13	.0000	.0000	.0000	.0001	.0009
	14	.0000	.0000	.0000	.0000	.0001

TABLE T5 A table of binomial probabilities for selected values of n and p (cont.)

n	x	.10	.20	p .30	.40	.50
15	0	.0352	.2059	.0047	.0005	.0000
	1	.1319	.3432	.0305	.0047	.0005
	2	.2309	.2669	.0916	.0219	.0032
	3	.2501	.1285	.1700	.0634	.0139
	4	.1876	.0428	.2186	.1268	.0417
	5	.1032	.0105	.2061	.1859	.0916
	6	.0430	.0019	.1472	.2066	.1527
	7	.0138	.0003	.0811	.1771	.1964
	8	.0035	.0000	.0348	.1181	.1964
	9	.0007	.0000	.0116	.0612	.1527
	10	.0001	.0000	.0030	.0245	.0916
	11	.0000	.0000	.0006	.0074	.0417
	12	.0000	.0000	.0001	.0016	.0139
	13	.0000	.0000	.0000	.0003	.0032
	14	.0000	.0000	.0000	.0000	.0005
	15	.0000	.0000	.0000	.0000	.0000

Example: With $n = 12$ and $p = 0.40$, $P(x = 2) = 0.0639$

TABLE T5   A table of Poisson probabilities for selected values of λ

x	0.5	1.0	1.5	2.0	λ 2.5	3.0	3.5	4.0	4.5	5.0
0	.6065	.3679	.2231	.1353	.0821	.0498	.0302	.0183	.0111	.0067
1	.3033	.3679	.3347	.2707	.2052	.1494	.1057	.0733	.0500	.0337
2	.0758	.1839	.2510	.2707	.2565	.2240	.1850	.1465	.1125	.0842
3	.0126	.0613	.1255	.1804	.2138	.2240	.2158	.1954	.1687	.1404
4	.0016	.0153	.0471	.0902	.1336	.1680	.1888	.1954	.1898	.1755
5	.0002	.0031	.0141	.0361	.0668	.1008	.1322	.1563	.1708	.1755
6	.0000	.0005	.0035	.0120	.0278	.0540	.0771	.1042	.1281	.1462
7	.0000	.0001	.0008	.0034	.0099	.0216	.0385	.0595	.0824	.1044
8	.0000	.0000	.0001	.0009	.0031	.0081	.0169	.0298	.0463	.0653
9	.0000	.0000	.0000	.0002	.0009	.0027	.0066	.0132	.0232	.0363
10	.0000	.0000	.0000	.0000	.0002	.0008	.0023	.0053	.0104	.0181
11	.0000	.0000	.0000	.0000	.0000	.0002	.0007	.0019	.0043	.0082
12	.0000	.0000	.0000	.0000	.0000	.0001	.0002	.0006	.0016	.0034
13	.0000	.0000	.0000	.0000	.0000	.0000	.0001	.0002	.0006	.0013
14	.0000	.0000	.0000	.0000	.0000	.0000	.0000	.0001	.0002	.0005
15	.0000	.0000	.0000	.0000	.0000	.0000	.0000	.0000	.0001	.0002

x	6	7	8	9	λ 10	11	12	13	14	15
0	.0025	.0009	.0003	.0001	.0000	.0000	.0000	.0000	.0000	.0000
1	.0149	.0064	.0027	.0011	.0005	.0002	.0001	.0000	.0000	.0000
2	.0446	.0223	.0107	.0050	.0023	.0010	.0004	.0002	.0001	.0000
3	.0892	.0521	.0286	.0150	.0076	.0037	.0018	.0008	.0004	.0002
4	.1339	.0912	.0573	.0337	.0189	.0102	.0053	.0027	.0013	.0006
5	.1606	.1277	.0916	.0607	.0378	.0224	.0127	.0070	.0037	.0019
6	.1606	.1490	.1221	.0911	.0631	.0411	.0255	.0152	.0087	.0048
7	.1377	.1490	.1396	.1171	.0901	.0646	.0437	.0281	.0174	.0104
8	.1033	.1304	.1396	.1318	.1126	.0888	.0655	.0457	.0304	.0194
9	.0688	.1014	.1241	.1318	.1251	.1085	.0874	.0661	.0473	.0324
10	.0413	.0710	.0993	.1186	.1251	.1194	.1048	.0859	.0663	.0486
11	.0225	.0452	.0722	.0970	.1137	.1194	.1144	.1015	.0844	.0663
12	.0113	.0264	.0481	.0728	.0948	.1094	.1144	.1099	.0984	.0829
13	.0052	.0142	.0296	.0504	.0729	.0926	.1056	.1099	.1060	.0956
14	.0022	.0071	.0169	.0324	.0521	.0728	.0905	.1021	.1060	.1024
15	.0009	.0033	.0090	.0194	.0347	.0534	.0724	.0885	.0989	.1024
16	.0003	.0014	.0045	.0109	.0217	.0367	.0543	.0719	.0866	.0960
17	.0001	.0006	.0021	.0058	.0128	.0237	.0383	.0550	.0713	.0847
18	.0000	.0002	.0009	.0029	.0071	.0145	.0256	.0397	.0554	.0706
19	.0000	.0001	.0004	.0014	.0037	.0084	.0161	.0272	.0409	.0557

TABLE T6   A table of Poisson probabilities for selected values of λ (cont.)

20	.0000	.0000	.0002	.0006	.0019	.0046	.0097	.0177	.0286	.0418
21	.0000	.0000	.0001	.0003	.0009	.0024	.0055	.0109	.0191	.0299
22	.0000	.0000	.0000	.0001	.0004	.0012	.0030	.0065	.0121	.0204
23	.0000	.0000	.0000	.0000	.0002	.0006	.0016	.0037	.0074	.0133
24	.0000	.0000	.0000	.0000	.0001	.0003	.0008	.0020	.0043	.0083
25	.0000	.0000	.0000	.0000	.0000	.0001	.0004	.0010	.0024	.0050
26	.0000	.0000	.0000	.0000	.0000	.0000	.0002	.0005	.0013	.0029
27	.0000	.0000	.0000	.0000	.0000	.0000	.0001	.0002	.0007	.0016
28	.0000	.0000	.0000	.0000	.0000	.0000	.0000	.0001	.0003	.0009
29	.0000	.0000	.0000	.0000	.0000	.0000	.0000	.0001	.0002	.0004
30	.0000	.0000	.0000	.0000	.0000	.0000	.0000	.0000	.0001	.0002
31	.0000	.0000	.0000	.0000	.0000	.0000	.0000	.0000	.0000	.0001
32	.0000	.0000	.0000	.0000	.0000	.0000	.0000	.0000	.0000	.0001

Example: With $\lambda = 3.0$, $P(x = 5) = 0.1008$

Index